The Excel Manual

Beverly Dretzke
Universty of Wisconsin - Eau Claire

SECOND EDITION

ELEMENTARY STATISTICS

Picturing the World

LARSON • FARBER

Prentice
Hall

Upper Saddle River, NJ 07458

Editor in Chief: Sally Yagan
Acquisitions Editor: Quincy McDonald
Associate Editor: Joanne Wendelken
Assistant Managing Editor: John Matthews
Production Editor: Donna Crilly
Supplement Cover Management/Design: Paul Gourhan
Manufacturing Buyer: Ilene Kahn

Prentice
Hall

© 2003 by Pearson Education
Pearson Education, Inc.
Upper Saddle River, NJ 07458

The author and publisher of this book have used their best efforts in preparing this book. These efforts include the development, research, and testing of the theories and programs to determine their effectiveness. The author and publisher make no warranty of any kind, expressed or implied, with regard to these programs or the documentation contained in this book. The author and publisher shall not be liable in any event for incidental or consequential damages in connection with, or arising out of, the furnishing, performance, or use of these programs.

Printed in the United States of America

10 9 8 7 6 5 4 3
ISBN 0-13-065948-7

Pearson Education Ltd., *London*
Pearson Education Australia Pty. Ltd., *Sydney*
Pearson Education Singapore, Pte. Ltd.
Pearson Education North Asia Ltd., *Hong Kong*
Pearson Education Canada, Inc., *Toronto*
Pearson Educacíon de Mexico, S.A. de C.V.
Pearson Education—Japan, *Tokyo*
Pearson Education Malaysia, Pte. Ltd.
Pearson Education, *Upper Saddle River, New Jersey*

▶ Contents:

Getting Started with Microsoft Excel

Overview

This manual is intended as a companion to Larson and Farber's *Elementary Statistics*. It presents instructions on how to use Microsoft Excel to carry out selected examples and exercises from *Elementary Statistics*.

The first section of the manual contains an introduction to Microsoft Excel and how to perform basic operations such as entering data, using formulas, saving worksheets, retrieving worksheets, and printing. All the screens pictured in this manual were obtained using the Office 2000 version of Microsoft Excel on a PC. You may notice slight differences if you are using a different version or a different computer.

Getting Started with the User Interface

GS 1.1	The Mouse

The mouse is a pointer device that allows you to move around the Excel worksheet and to select specific locations and objects. There are four main mouse operations: Select, click, double-click, and right-click.

1. To **select** generally means to move the mouse pointer so that the white arrow is pointing at or is positioned directly over an object. You will often **select** commands in the standard toolbar located near the top of the screen. Some of the more familiar of these commands are open, save, and print.

2. To **click** means to press down on the left button of the mouse. You will frequently select cells of the worksheet and commands by "clicking" the left button.

3. To **double-click** means to press the left mouse button twice in rapid succession.

4. To **right-click** means to press down on the right button of the mouse. A right-click is often used to display special shortcut menus.

GS 1.2	The Excel Worksheet

The figure shown below presents the Office 2000 version of a blank Excel worksheet. Important parts of the worksheet are labeled.

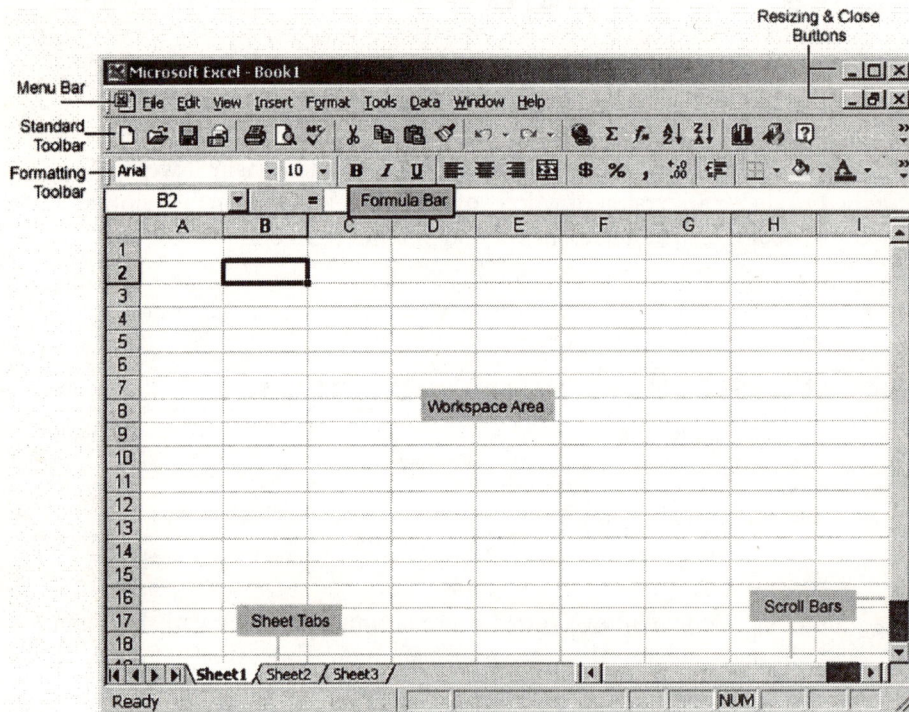

GS 1.3	Menu Conventions

Excel uses standard conventions for all menus. For example, the Menu bar contains the commands File, Edit, View, etc. Selecting one of these commands will "drop down" a menu. The Edit menu is displayed at the top of the next page.

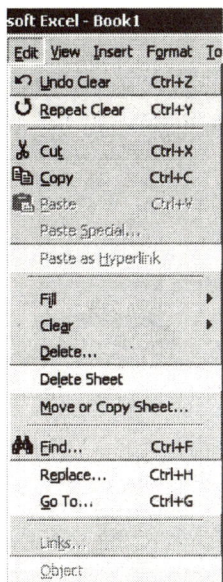

Icons to the left of the Cut, Copy, Paste, and Find commands indicate toolbar buttons that are equivalent to the menu choices.

Keyboard shortcuts are displayed to the right of the commands. For example, Ctrl+X is a keyboard shortcut for Cut.

The triangular markers to the right of Fill and Clear indicate that selection of these commands will result in a second menu of choices.

Selection of commands that are followed by an ellipsis (e.g., Paste Special… and Delete…) will result in the display of a dialog box that usually must be responded to in some way in order for the command to be executed.

The menus found in other locations of the Excel worksheet will operate in the same way.

GS 1.4   Dialog Boxes

Many of the statistical analysis procedures that are presented in this manual are associated with commands that are followed by dialog boxes. Dialog boxes usually require that you select from alternatives that are presented or that you enter your choices.

For example, if you select **Insert → Function**, a dialog box like the one shown below will appear. You are required to select both a Function category and a Function name. You make your selections by clicking on them.

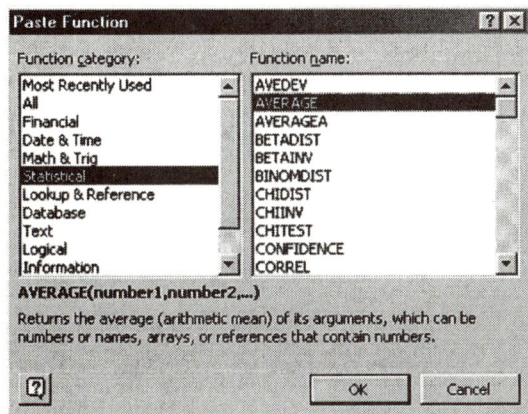

When you click the OK button at the bottom of the dialog box, another dialog box will often be displayed that asks you to provide information regarding location of the data in the Excel worksheet

Getting Started with Opening Files

GS 2.1	Opening a New Workbook

When you start Excel, the screen will open to **Sheet 1** of **Book 1**. Sheet names appear on tabs at the bottom of the screen. The name "Book 1" will appear in the top left corner.

If you are already working in Excel and have finished the analyses for one problem and would like to open a new book for another problem, follow these steps: First, at the top of the screen, click **File → New**. Next, click **OK** in the New dialog box. If you were previously working in Book1, the new worksheet will be given the default name Book 2.

The names of books opened during an Excel work session will be displayed at the bottom of the Window menu. To return to one of these books, click **Window** and then click the book name.

GS 2.2	Opening a File That Has Already Been Created

To open a file that you or someone else has already created, click on **File →Open**. A list of file locations will appear. Select the location by clicking on it. Many of the data files that are presented in your statistics textbook are available on the 3 ½ floppy disk that accompanies this manual. To open any of these files, you will select **3½ Floppy (A:)**.

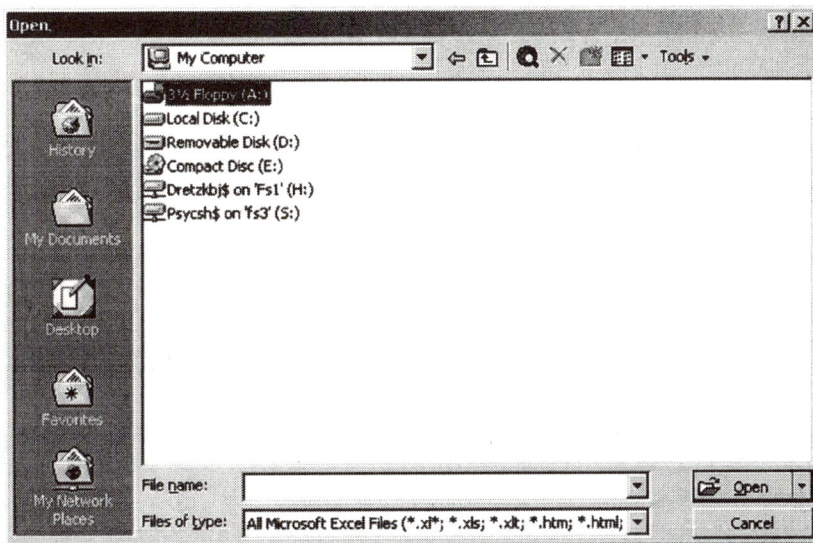

After you click on 3½ Floppy (A:), a list of folders and files available on the 3½ floppy disk will appear. You will need to select the folder or file you want by clicking on it. If you have selected a folder, another screen will appear with a list of files contained in the folder. Click on the name of the file that you would like to open.

Getting Started with Entering Information

GS 3.1	Cell Addresses

Columns of the worksheet are identified by letters of the alphabet and rows are identified by numbers. The cell address A1 refers to the cell located in column A row 1. The dark outline around a cell means that it is "active" and is ready to receive information. In the figure shown below, cell C1 is ready to receive information. You can also see C1 in the **Name Box** to the left of the **Formula Bar**. You can move to different cells of the worksheet by using the mouse pointer and clicking on a cell. You can also press [**Tab**] to move to the right or left, or you can use the arrow keys on the keyboard.

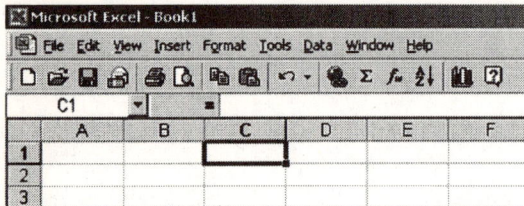

You can also activate a **range** of cells. To activate a range of cells, first click in the top cell and drag down and across (or click in the bottom cell and drag up and across). The range of cells highlighted in the figure below is designated B2:D6.

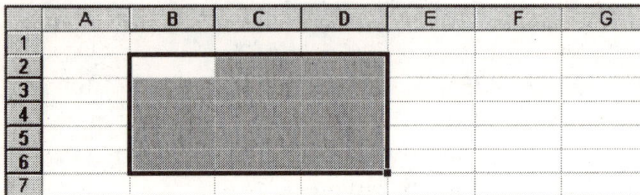

GS 3.2	Types of Information

Three types of information may be entered into an Excel worksheet.

1. **Text**. The term "text" refers to alphabetic characters or a combination of alphabetic characters and numbers, sometimes called "alphanumeric." The figure provides an

example of an entry comprised solely of alphabetic characters (cell A1) and an entry comprised of a combination of alphabetic characters and numbers (cell B1).

	A	B	C	D	E	F	G
1	Sue Clark	25 years					
2							
3							

2. **Numeric**. Any cell entry comprised completely of numbers falls into the "numeric" category.

3. **Formulas**. Formulas are a convenient way to perform mathematical operations on numbers already entered into the worksheet. Specific instructions are provided in this manual for problems that require the use of formulas.

GS 3.3	Entering Information

To enter information into a cell of the worksheet, first activate the cell. Then key in the desired information and press [**Enter**]. Pressing the [Enter] key moves you down to the next cell in that column. The information shown below was entered as follows:

1. Click in cell A1. Key in **1**. Press [**Enter**].

2. Key in **2**. Press [**Enter**].

3. Key in **3**. Press [**Enter**].

	A	B	C	D	E	F	G
1	1						
2	2						
3	3						
4							
5							

GS 3.4	Using Formulas

When you want to enter a formula, begin the cell entry with an equal sign (=). The arithmetic operators are displayed at the top of the next page.

Arithmetic operator	Meaning	Example
+	Addition	3+2
-	Subtraction	3-2
*	Multiplication	3*2
/	Division	3/2
^	Exponentiation	3^2

Numbers, cell addresses, and functions can be used in formulas. For example, to sum the contents of cells A1 and B1, you can use the formula =A1+B1. To divide this sum by 2, you can use the formula =(A1+B1)/2. Note that Excel carries out expressions in parentheses first and then uses the results to complete the calculations of the formula. Formulas will sometimes not produce the desired results because parentheses were necessary but were not used.

Getting Started with Changing Information

GS 4.1	Editing Information in the Cells

There are several ways that you can edit information that has already been entered into a cell.

1. If you have not completed the entry, you can simply backspace and start over. Clicking on the red X to the left of the Formula Bar will also delete an incomplete cell entry.

2. If you have already completed the entry and another cell is activated, you can click on the cell you want to edit and then press either [**Delete**] or [**Backspace**] to clear the contents of the cell.

3. If you want to edit part of the information in a cell instead of deleting all of it, follow the instructions provided in the example.

- Let's say that you wanted to enter 1234 in cell A1 but instead entered 124. Return to cell **A1** to make it the active cell by either clicking on it with the mouse or by using the arrow keys.

A1	▼	▪	124				
A	**B**	**C**	**D**	**E**	**F**	**G**	
1	124						
2							

- You will see A1 in the Name Box and 124 in the Formula Bar. Click between 2 and 4 in the Formula Bar so that the **I-beam** is positioned there. Enter the number 3 and press [**Enter**].

GS 4.2	Copying Information

To copy the information in one cell to another cell, follow these steps:

- First click on the source cell. Then, at the top of the screen, click **Edit →Copy**.

- Click on the target cell where you want the information to be placed. Then, at the top of the screen, click **Edit → Paste**.

To copy a range of cells to another location in the worksheet, follow these steps:

- First click and drag over the range of cells that you want to copy so that they are highlighted. Then, at the top of the screen, click **Edit →Copy**.

- Click in the topmost cell of the target location. Then, at the top of the screen, click **Edit →Paste**.

To copy the contents of one cell to a range of cells follow these steps:

- Let's say that you have entered a formula in cell C1 that adds the contents of cells A1 and B1 and you would like to copy this formula to cells C2 and C3 so that C2 will contain the sum of A2 and B2 and cell C3 will contain the sum of A3 and B3.

- First click on cell C1 to make it the active cell. You will see =A1+B1 in the Formula Bar.

C1			=	=A1+B1			
	A	B	C	D	E	F	G
1	1	1	2				
2	2	2					
3	3	3					

- At the top of the screen, click **Edit → Copy**.

- Highlight cells C2 and C3 by clicking and dragging over them.

	A	B	C	D	E	F	G
1	1	1	2				
2	2	2					
3	3	3					
4							

- At the top of the screen, click **Edit → Paste**. The sums should now be displayed in cells C2 and C3.

C2			=	=A2+B2			
	A	B	C	D	E	F	G
1	1	1	2				
2	2	2	4				
3	3	3	6				
4							

GS 4.3	Moving Information

If you would like to move the contents of one cell from one location to another in the worksheet, follow these steps:

- Click on the cell containing the information that you would like to move.

- At the top of the screen, click **Edit →Cut**.

- Click on the target cell where you want the information to be placed.

- At the top of the screen, click **Edit → Paste**.

If you would like to move the contents of a range of cells to a different location in the worksheet, follow these steps:

- Click and drag over the range of cells that you would like to move so that it is highlighted.

- At the top of the screen, click **Edit →Cut**.

- Click the topmost cell of the new location. (It is not necessary to click and drag over the entire range of the new location.)

- At the top of the screen, click **Edit → Paste**.

*If you make a mistake, just click **Edit → Undo**.*

GS 4.4	Changing the Column Width

There are a couple of different ways that you can use to change the column width. Only one way will be described here. Output from the Descriptive Statistics data analysis tool will be used as an example. As you can see in the output displayed below, many of the labels in column A can only be partially viewed because the column width is too narrow.

	A	B
1	*Test Score*	
2		
3	Mean	23.89474
4	Standard E	0.752752
5	Median	24
6	Mode	21
7	Standard C	3.281171
8	Sample Va	10.76608

Position the mouse pointer directly on the vertical line between A and B in the letter row at the top of the worksheet — A   B —so that it turns into a black plus sign.

Click and drag to the right until you can read all the output labels. (You can also click and drag to the left to make columns narrower.) After adjusting the column width, your output should appear similar to the output shown below.

	A	B
1	Test Score	
2		
3	Mean	23.89474
4	Standard Error	0.752752
5	Median	24
6	Mode	21
7	Standard Deviation	3.281171
8	Sample Variance	10.76608

Getting Started with Sorting Information

GS 5.1   Sorting a Single Column of Information

Let's say that you have entered "Score" in cell A1 and four numbers directly below it and that you would like to sort the numbers in ascending order.

	A	B	C	D	E	F	G
1	Score						
2	15						
3	79						
4	18						
5	2						

- Click and drag from cell A1 to cell A5 so that the range of cells is highlighted.

You could also click directly on A *in the letter row at the top of the worksheet. This will result in all cells of column A being highlighted.*

	A	B	C	D	E	F	G
1	Score						
2	15						
3	79						
4	18						
5	2						
6							

- At the top of the screen, click **Data** → **Sort**.

- In the Sort dialog box that appears, you are given the choice of sorting the information in column A in either ascending or descending order. The ascending order has already been selected. Header row has also been selected. This means that the "Score" header will stay in cell A1 and will not be included in the sort. Click **OK** at the bottom of the dialog box.

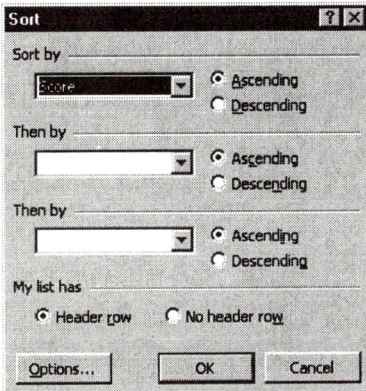

The cells in column A should now be sorted in ascending order as shown below.

GS 5.2	Sorting Multiple Columns of Information

Your Excel data files will frequently contain multiple columns of information. When you sort multiple columns at the same time, Excel provides a number of options.

Let's say that you have a data file that contains the information shown below and that you would like to sort the file by GPA in descending order.

	A	B	C	D	E	F	G
1	Score	Age	Major	GPA			
2	2	19	Music	3.1			
3	15	19	History	2.4			
4	18	22	English	2.7			
5	79	20	English	3.7			

- Click and drag from A1 down and across to D5 so that the entire range of cells is highlighted.

	A	B	C	D	E	F	G
1	Score	Age	Major	GPA			
2	2	19	Music	3.1			
3	15	19	History	2.4			
4	18	22	English	2.7			
5	79	20	English	3.7			
6							

- At the top of the screen, click **Data** → **Sort**.

- In the Sort dialog box that appears, you are given the option of sorting the data by three different variables. You want to sort only by GPA in descending order. Click the down arrow to the right of the Sort by window until you see GPA and click on **GPA** to select it. Then click the button to the left of **Descending** so that a black dot appears there. You want the variable labels to stay in row 1, so **Header row** should be selected. Click **OK**.

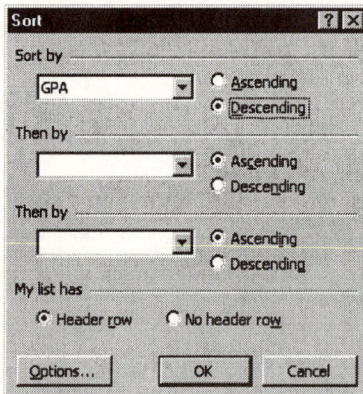

Sort

Sort by
GPA
○ Ascending
● Descending

Then by
● Ascending
○ Descending

Then by
● Ascending
○ Descending

My list has
● Header row   ○ No header row

Options...   OK   Cancel

The sorted data file is shown below.

	A	B	C	D	E	F	G
1	Score	Age	Major	GPA			
2	79	20	English	3.7			
3	2	19	Music	3.1			
4	18	22	English	2.7			
5	15	19	History	2.4			
6							

Getting Started with Saving Information

GS 6.1	Saving Files

To save a newly created file for the first time, click **File → Save** at the top of the screen. A Save As dialog box will appear. You will need to select the location for saving the file by clicking on it. In the dialog box shown below, the 3½ Floppy has been selected.

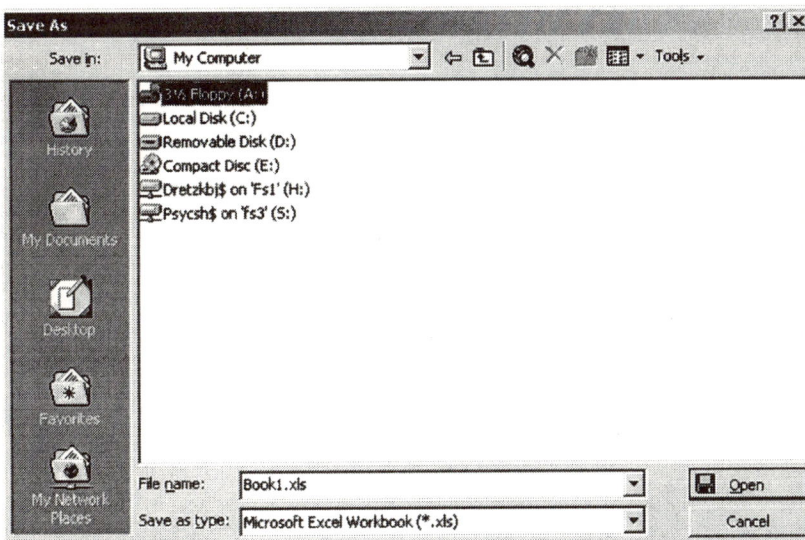

Save As ? ×

Save in: My Computer

- 3½ Floppy (A:)
- Local Disk (C:)
- Removable Disk (D:)
- Compact Disc (E:)
- Dretzkbj$ on 'Fs1' (H:)
- Psycsh$ on 'Fs3' (S:)

File name: Book1.xls   Open

Save as type: Microsoft Excel Workbook (*.xls)   Cancel

The default file name, displayed in the File name window, is **Book1.xls**. It is highly recommended you replace the default name with a name that is more descriptive. It is also highly recommended that you use the **xls** extension for all your Excel files.

Once you have saved a file, clicking **File → Save** will result in the file being saved in the same location under the same file name. No dialog box will appear. If you would like to save the file in a different location, you will need to click **File → Save As**.

GS 6.2	Naming Files

Windows 2000 and Mac versions of Excel will allow file names to have around 200 characters. The extension can have up to three characters. You will find that long, descriptive names will be easier to work with than really short names. For example, if a file contains data that was collected in a survey of Milwaukee residents, you may want to name the file **Milwaukee survey.xls**.

Several symbols cannot be used in file names. These include: forward slash (/), backslash (\), greater-than sign (>), less-than sign (<), asterisk (*), question mark (?), quotation mark ("), pipe symbol (|), color (:), and semicolon (;).

Getting Started with Printing Information

GS 7.1	Printing Worksheets

To print a worksheet, click **File → Print**. The Print dialog box will appear.

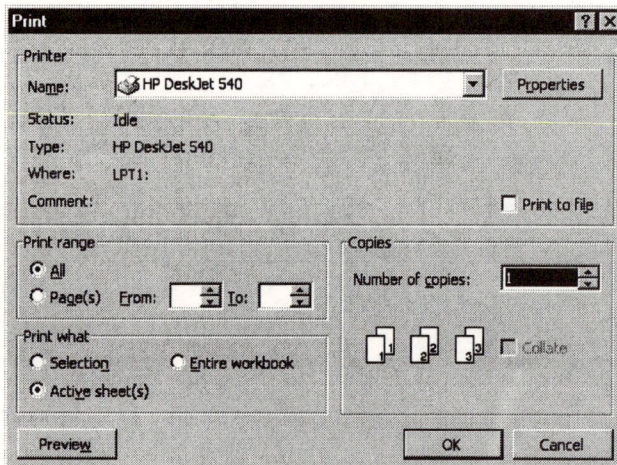

Under Print range, you will usually select **All**, and under Print what, you will usually select **Active sheet(s)**. The default number of copies is 1, but you can increase this if you need more copies. When the Print dialog box has been completed as you would like, click **OK**.

GS 7.2	Page Setup

Excel provides a number of page setup options for printing worksheets. To access these options, click **File → Page Setup**. Under **Page**, you may want to select the Landscape orientation for worksheets that have several columns of data. Under **Sheet**, you may want to select Gridlines.

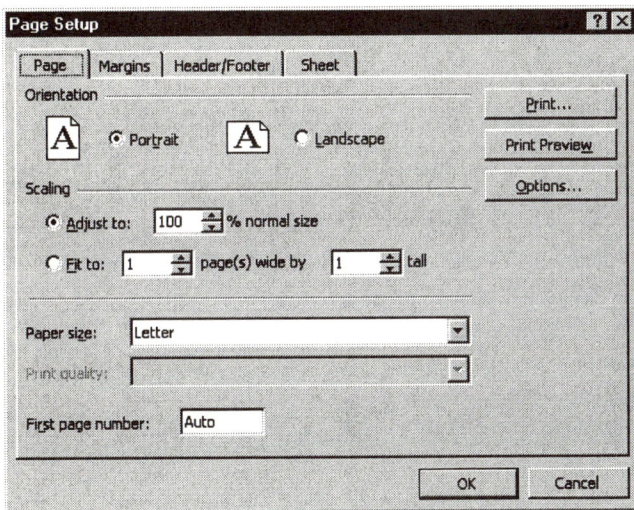

Getting Started with Add-ins

GS 8.1	Loading Excel's Analysis Toolpak

The Analysis Tookpak is an Excel Add-In that may not necessarily be loaded. If it does not appear at the bottom of the Tools menu, then click on **Add-Ins** in the Tools menu to get the dialog box shown at the top of the next page.

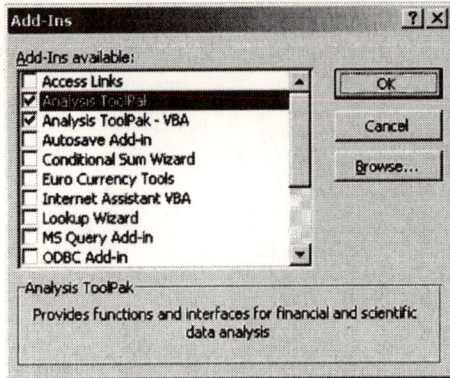

Click in the box to the left of **Analysis ToolPak** to place a checkmark there. Then click **OK**. The ToolPak will load and will be listed at the bottom of the Tools menu as shown below.

| GS 8.2 | Loading the PHStat2 Add-In |

PHStat2 is a Prentice Hall statistical add-in that is included on the CD-ROM that accompanies your statistics textbook. The instructions that are given here also appear in the PHStat readme file.

To use the Prentice Hall PHStat2 Microsoft Excel add-in, you first need to run the setup program (Setup.exe) located in the PHStat2 folder on your textbook CD-ROM. The setup program will install the PHStat2 program files to your system and add icons on your Desktop and Start Menu for PHStat2. Depending on the age of your Windows system files, some Windows system files may be updated during the setup process as well. Note that PHStat2 is compatible with Microsoft Excel 97 and Microsoft Excel 2000. PHStat2 is not compatible with Microsoft Excel 95.

During the Setup program you will have the opportunity to specify:

a.   The directory into which to copy the PHStat2 files (default is \Program Files\Prentice Hall\PHStat2).

b.   The name of the Start Programs folder to contain the PHStat2 icon.

To begin using the Prentice Hall PHStat2 Microsoft Excel add-in, click the appropriate Start Menu or Desktop icon for PHStat2 that was added to your system during the setup process.

When a new, blank Excel worksheet appears, check the Tools menu to make sure that both the **Analysis ToolPak** and **Analysis ToolPak–VBA** have been checked.

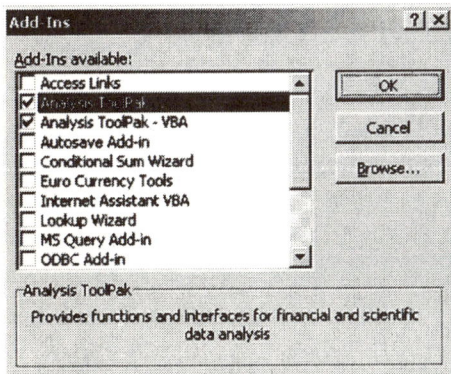

Introduction to Statistics

Technology Lab

| ▶ Example (pg. 22) | Generating a List of Random Numbers |

You will be generating a list of 15 random numbers between 1 and 167 to use in selecting a random sample of 15 cars assembled at an auto plant.

If the PHStat2 add-in has not been loaded, you will need to load it before continuing. Follow the instructions in Section GS 8.2.

1. Open a new, blank Excel worksheet.

2. At the top of the screen, select **PHStat → Sampling → Random Sample Generation**.

3. Complete the Random Sample Generator dialog box as shown below. A sample of 15 cars will be randomly selected from a population of 167. The topmost cell in the output will contain the title "Car #." Click **OK**.

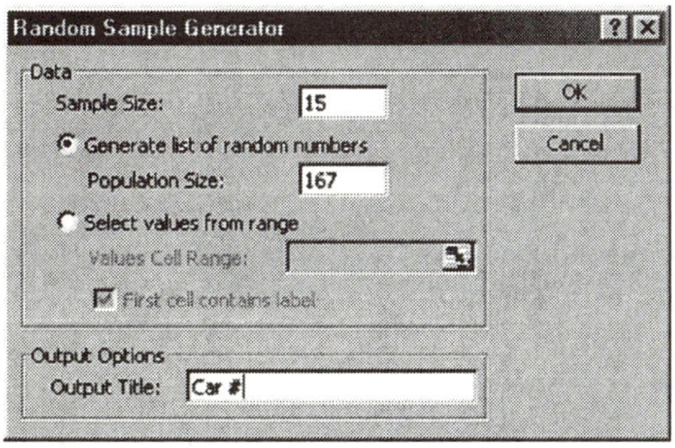

The output is displayed in a new worksheet named "RandomNumbers." Because the numbers were generated randomly, it is not likely that your output will be exactly the same.

	A	B	C	D	E	F	G
1	Car #						
2	119						
3	20						
4	9						
5	35						
6	82						
7	45						
8	39						
9	148						
10	127						
11	58						
12	47						
13	156						
14	14						
15	34						
16	123						

◀

| ▶ Exercise 1 (pg. 23) | Generating a List of Random Numbers for a CPA |

You will be generating a list of 8 random numbers between 1 and 74 to use in selecting a random sample of accounts. You will also be ordering the generated list from lowest to highest.

If the PHStat2 add-in has not been loaded, you will need to load it before continuing. Follow the instructions in Section GS 8.2.

1. Open a new, blank Excel worksheet.

2. At the top of the screen, select **PHStat → Sampling → Random Sample Generation**.

3. Complete the Random Sample Generator dialog box as shown below. A sample of eight accounts will be randomly selected from a population of 74. "Account #" will appear in the top cell of the output. Click **OK**.

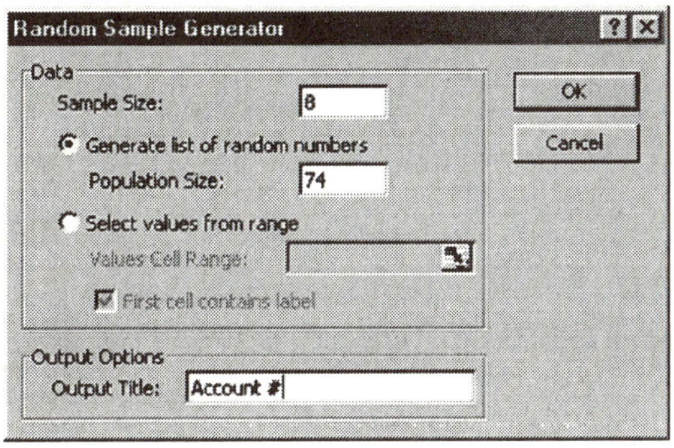

```
┌──────────────────────────────────────────────────┐
│ Random Sample Generator                    [?][X] │
│ ┌─Data────────────────────────┐  ┌───────────┐    │
│ │  Sample Size:        [8    ] │  │    OK     │    │
│ │  (•) Generate list of random numbers          │
│ │      Population Size:  [74  ] │  │  Cancel   │    │
│ │  ( ) Select values from range │  └───────────┘    │
│ │      Values Cell Range: [      ] [▦] │             │
│ │      [✓] First cell contains label    │             │
│ └──────────────────────────────┘                  │
│ ┌─Output Options──────────────────────┐           │
│ │  Output Title:  [Account #|        ] │           │
│ └─────────────────────────────────────┘           │
└──────────────────────────────────────────────────┘
```

4. Sort the numbers in ascending order.

For instructions on how to sort, refer to Sections GS 5.1 and GS 5.2.

The sorted set of eight random numbers is displayed below. Because the numbers were generated randomly, it is not likely that your output will be exactly the same.

	A	B	C	D	E	F	G
1	Account #						
2	13						
3	20						
4	32						
5	43						
6	57						
7	58						
8	59						
9	68						

◀

▶ **Exercise 2 (pg. 23)** | Generating a List of Random Numbers for a Quality Control Department

---

If the PHStat2 add-in has not been loaded, you will need to load it before continuing. Follow the instructions in Section GS 8.2.

---

1. Open a new Excel worksheet.

2. At the top of the screen, select **PHStat** → **Sampling** → **Random Sample Generation**.

3. Complete the Random Sample Generator dialog box as shown below. A sample of 20 batteries will be randomly selected from a population of 200. "Battery #" will appear in the top cell of the generated output. Click **OK**.

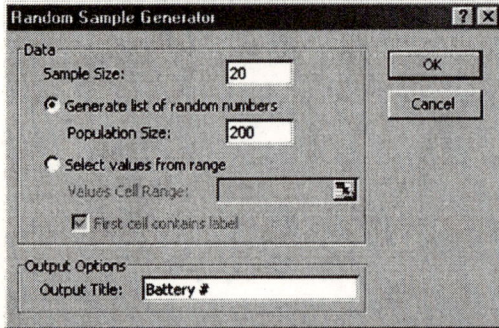

```
┌─ Random Sample Generator ──────────────── [?][X]┐
│ ┌─Data────────────────────────┐  ┌──────────┐   │
│ │  Sample Size:        [20  ] │  │    OK    │   │
│ │  ⦿ Generate list of random  │  └──────────┘   │
│ │     numbers                 │  ┌──────────┐   │
│ │  Population Size:    [200 ] │  │  Cancel  │   │
│ │  ○ Select values from range │  └──────────┘   │
│ │  Values Cell Range: [    ]⬛ │                 │
│ │     ☑ First cell contains   │                 │
│ │         label               │                 │
│ └─────────────────────────────┘                 │
│ ┌─Output Options──────────────────────────────┐ │
│ │  Output Title:  [Battery #                ] │ │
│ └─────────────────────────────────────────────┘ │
└──────────────────────────────────────────────────┘
```

4. Sort the numbers in ascending order.

For instructions on how to sort, see Sections GS 5.1 and GS 5.2.

The sorted set of 20 random numbers is displayed below. Because the numbers were generated randomly, it is unlike that your output will be exactly the same.

	A	B	C	D	E	F	G
1	Battery #						
2	1						
3	5						
4	11						
5	50						
6	58						
7	59						
8	62						
9	65						
10	76						
11	85						
12	91						
13	93						
14	111						
15	117						
16	148						
17	155						
18	165						
19	168						
20	182						
21	186						

◀

► Exercise 3 (pg. 23)	Generating Three Random Samples from a Population of Ten Digits

You will be generating three random samples of five digits each from the population: 0, 1, 2, 3, 4, 5, 6, 7, 8, 9. You will also compute the average of each sample.

> *If the PHStat2 add-in has not been loaded, you will need to load it before continuing. Follow the instructions in Section GS 8.2.*

1. Open a new Excel worksheet. Enter the numbers 0 through 9 in column A.

	A	B	C	D	E	F	G
1	0						
2	1						
3	2						
4	3						
5	4						
6	5						
7	6						
8	7						
9	8						
10	9						

2. Compute the average using the AVERAGE function. To do this, first click in the cell immediately below the last number in the population—cell A11. Then, at the top of the screen click **Insert → Function**.

3. Under Function category, select **Statistical**. Under Function name, select **AVERAGE**. Click **OK**.

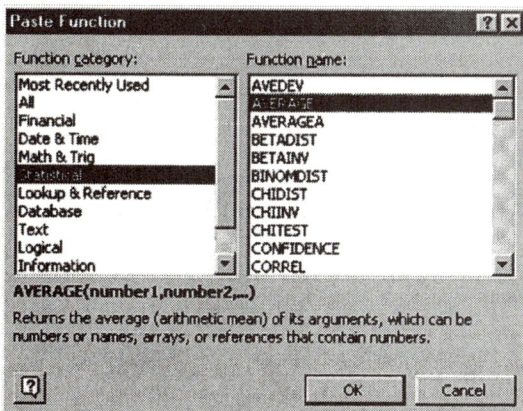

4. Complete the AVERAGE dialog box as shown below. Click **OK**.

A quick way to enter the range is to first click in the Number 1 field in the dialog box. Then click on cell A1 in the worksheet and drag down to cell A10.

AVERAGE
Number1 [A1:A10]   = {0;1;2;3;4;5;6;7;8;9
Number2 [   ]   = number

= 4.5
Returns the average (arithmetic mean) of its arguments, which can be numbers or names, arrays, or references that contain numbers.
Number1: number1,number2,... are 1 to 30 numeric arguments for which you want the average.

Formula result =4.5   [OK]   [Cancel]

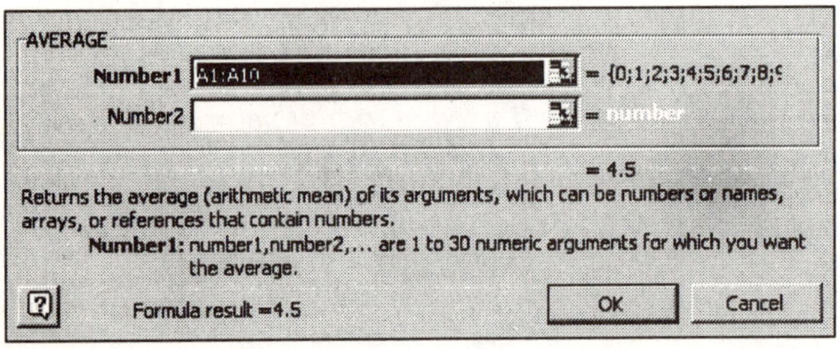

The average, 4.5, is displayed in cell A11 of the worksheet.

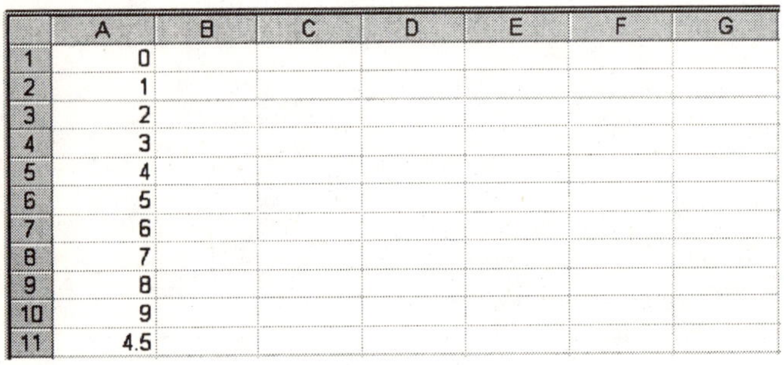

	A	B	C	D	E	F	G
1	0						
2	1						
3	2						
4	3						
5	4						
6	5						
7	6						
8	7						
9	8						
10	9						
11	4.5						

5. At the top of the screen, select **PHStat → Sampling→ Random Sample Generation**.

6. Complete the Random Sample Generator dialog box as shown at the top of the next page. Five numbers will be randomly selected from the numbers displayed in cells A1 through A10 in the worksheet. Cell A1 does not contain a label. "Sample 1" will appear in the top cell of the generated output. Click **OK**.

A quick way to enter the range is to first click in the Values Cell Range field of the dialog box. Then click on cell A1 in the worksheet and drag down to cell A10. Be careful that you do not include cell A11 in the range.

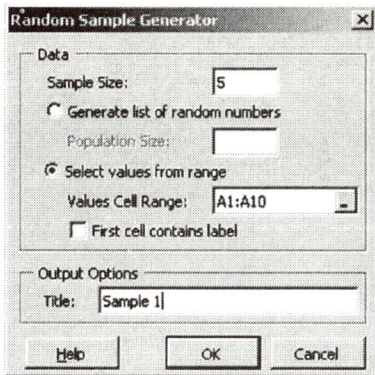

The random sample of five digits is placed in a sheet named "RandomNumbers." Because the numbers were generated randomly, it is not likely that your output will be exactly the same.

	A	B	C	D	E	F	G
1	Sample 1						
2	6						
3	5						
4	0						
5	9						
6	2						

7. Use the AVERAGE function to find the average. To do this, first click in the cell immediately below the last number in the sample—cell A7. Then, at the top of the screen, click **Insert →Function**.

8. Under Function category, select **Statistical**. Under Function name, select **AVERAGE**. Complete the AVERAGE dialog box as shown below. Click **OK**.

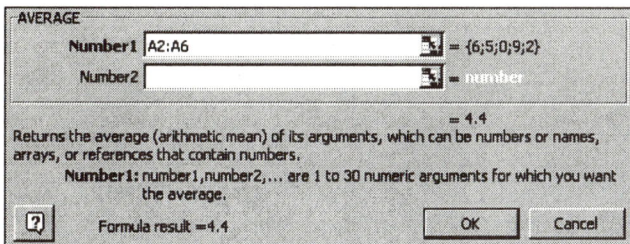

9. The average of the sample, 4.4, is now displayed in cell A7 of the worksheet. Repeat steps 6-9 to generate two more samples and compute the averages.

To return to the sheet that contains the population values, click on the Sheet1 tab located near the bottom of your screen.

	A	B	C	D	E	F	G
1	Sample 1						
2	6						
3	5						
4	0						
5	9						
6	2						
7	22						
8	4.4						

◄

► Exercise 5 (pg. 23)	Simulating Rolling a Six-Sided Die 60 Times

You will be simulating rolling a six-sided die 60 times and making a tally of the results.

If the PHStat2 add-in has not been loaded, you will need to load it before continuing. Follow the instructions in Section GS 8.2.

1. Open a new Excel worksheet and enter the digits 1 through 6 in column A. These digits represent the possible outcomes of rolling a die.

	A	B	C	D	E	F	G
1	1						
2	2						
3	3						
4	4						
5	5						
6	6						

2. At the top of the screen, click **Tools → Data Analysis**. Select **Sampling** and click **OK**.

If Data Analysis does not appear as a choice in the Tools menu, you will need to load the Microsoft Excel Analysis ToolPak add-in. Follow the procedure in Section GS 8.1 before continuing.

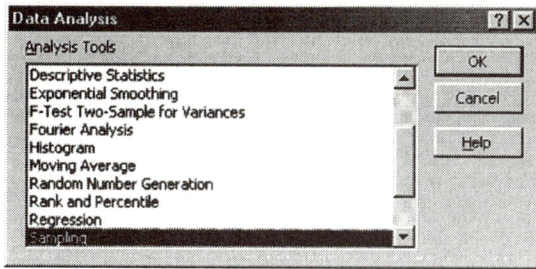

3. Complete the Sampling dialog box as shown below. Click **OK**.

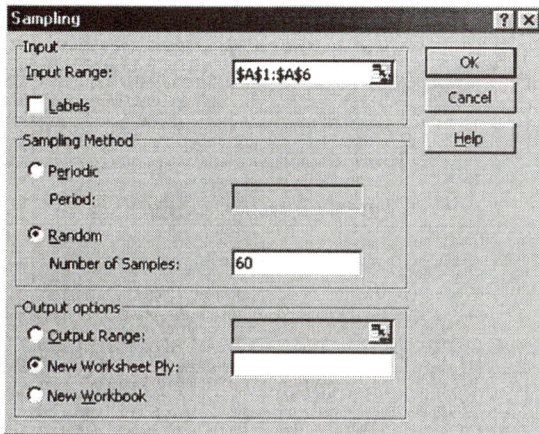

The output will be displayed in a new worksheet. The first 7 entries are shown below. Because the numbers were generated randomly, it is not likely that your output will be exactly the same.

	A	B	C	D	E	F	G
1	5						
2	3						
3	5						
4	6						
5	4						
6	5						
7	6						

4. To make a tally of the results, first select **PHStat → Descriptive Statistics → One-Way Tables & Charts**. Then complete the One-Way Tables & Charts dialog box as shown below. The Type of Data is "Raw Categorical Data." The data are located in cells A1 through A60. There is no label in the first cell (A1). The output will be given the title "Rolling a Die." A bar chart will be part of the output. Click **OK**.

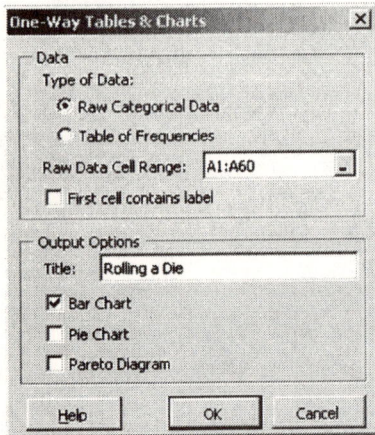

The bar chart is displayed in a sheet named "Bar Chart."

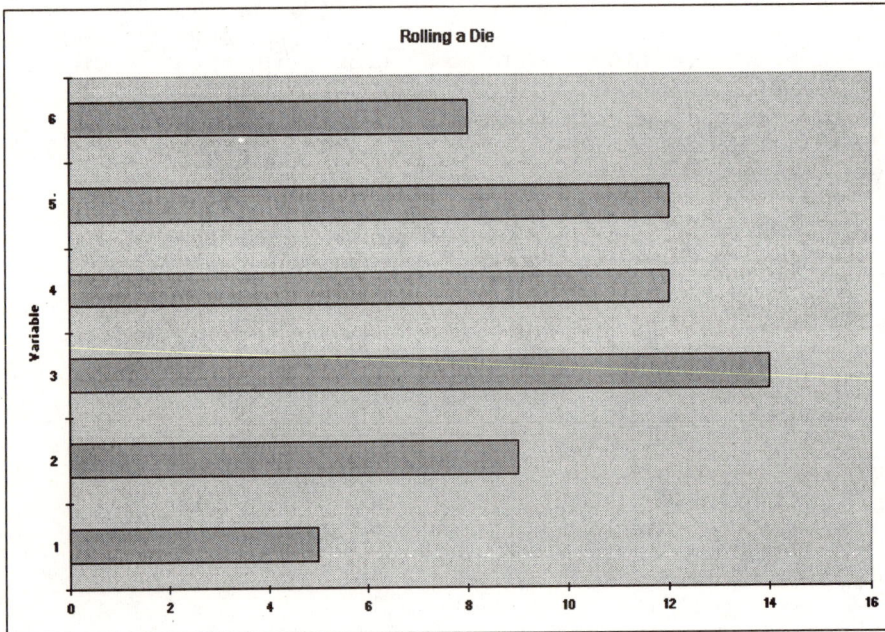

5. Click on the **OneWay Table** sheet tab at the bottom of the screen. This sheet contains a frequency distribution of the outcomes.

	A	B
1	Rolling a Die	
2		
3	Count of Variable	
4	Variable ▾	Total
5	1	5
6	2	9
7	3	14
8	4	12
9	5	12
10	6	8
11	Grand Total	60

6. Click on the **DataCopy** sheet tab at the bottom of the screen. This sheet contains a copy of the generated data set with the word "Variable" inserted in cell A1.

◀

Descriptive Statistics

Section 2.1

▶ Example 7 (pg. 40)	Constructing Histograms

You will be constructing a histogram for the frequency distribution of the Internet data in Example 2 on page 35.

1. Open worksheet "internet" in the Chapter 2 folder. These data represent the number of minutes 50 Internet subscribers spent during their most recent Internet session. You will use Data Analysis located in the Tools menu to construct a histogram for these data.

If Data Analysis does not appear as a choice in the Tools menu, you will need to load the Microsoft Excel Analysis ToolPak add-in. Follow the procedure in Section GS 8.1 before continuing.

---

2. Excel's histogram procedure uses grouped data to generate a frequency distribution and a frequency histogram. The procedure requires that you indicate a "bin" for each class. The number that you specify for each bin is actually the upper limit of the class. The upper limits for the Internet data are given on page 33 of your text and are based on a class width of 12. You see that 18 is the upper limit for the first class, 30 is the upper limit for the second class, and so on. The bin for the first class will contain a count of all observations less than or equal to 18, the bin for the

second class will contain a count of all observations between 19 and 30, and so on. Enter **Bin** in cell C1 and key in the upper limits in column C as shown below.

	A	B	C	D	E	F	G
1	Internet subscribers	Bin					
2	50		18				
3	40		30				
4	41		42				
5	17		54				
6	11		66				
7	7		78				
8	22		90				

3.   Click **Tools → Data Analysis**. Select **Histogram**, and click **OK**.

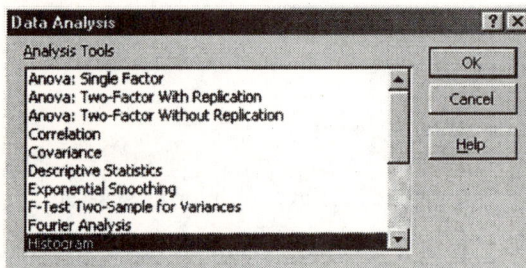

4.   Complete the Histogram dialog box as shown below and click **OK**.

To enter the input and bin ranges quickly, follow these steps. First click in the Input Range field of the dialog box. Then, in the worksheet, click and drag from cell A1 to cell A51. You will then see $A$1:$A$51 displayed in the Input Range field. Next click in the Bin Range field. Then click and drag from cell C1 through cell C8 in the worksheet. You will see $C$1:$C$8 displayed in the Bin Range field.

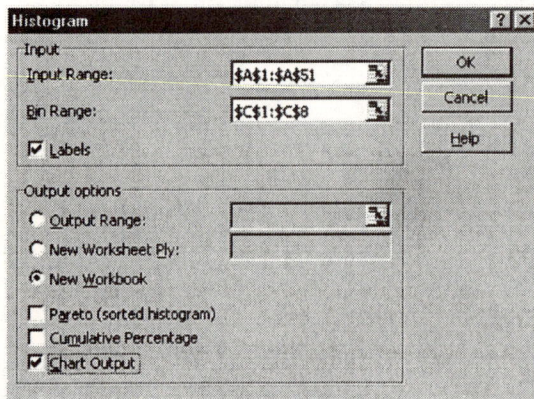

Note the checkmark in the box to the left of Labels. In your worksheet, "Internet subscribers" appears in cell A1, and "Bin" appears in cell C1. Because you included these cells in the Input Range and Bin Range, respectively, you need to let Excel know that these cells contain labels rather than data. Otherwise Excel will attempt to use the information in these cells when constructing the frequency distribution and histogram.

You should see output similar to the output displayed below.

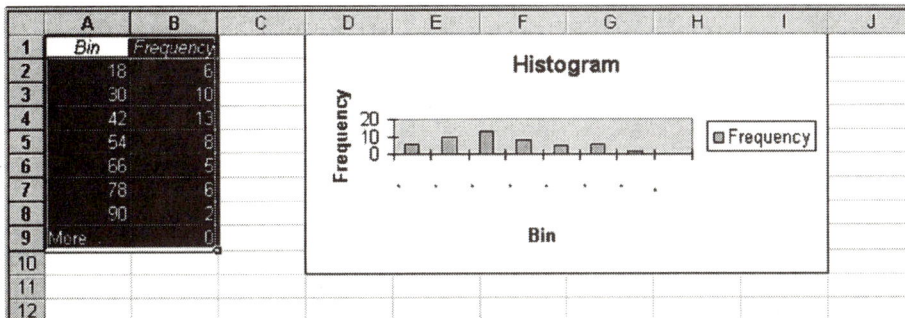

If you want to change the height and width of the chart, begin by clicking anywhere within the figure. Handles appear along the perimeter. You can change the shape of the figure by clicking on a handle and dragging it. You can make the figure small and short or you can make it tall and wide. You can also click within the figure and drag it to move it to a different location on the worksheet.

---

5. You will now follow steps to modify the histogram so that it is displayed in a more informative and attractive manner. First, make the chart taller so that it is easier to read. To do this, first click within the figure near a border. Black square handles appear. Click on the center handle on the bottom border of the figure and drag it down a few rows.

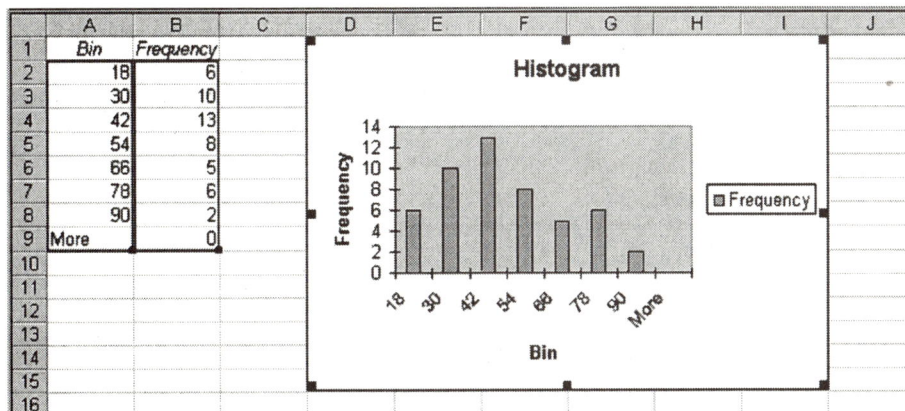

6. Next remove the space between the vertical bars. **Right click** on one of the vertical bars. Select **Format Data Series** from the shortcut menu that appears.

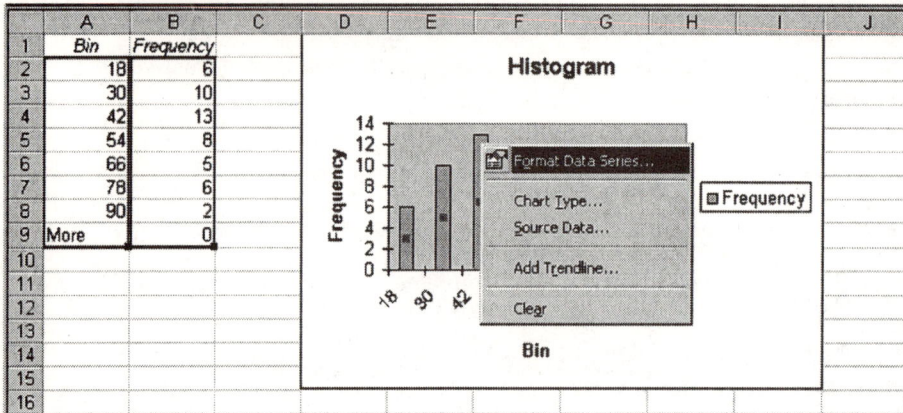

7. Click on the **Options** tab at the top of the Format Data Series dialog box. Change the value in the Gap width box to 0. Click **OK**.

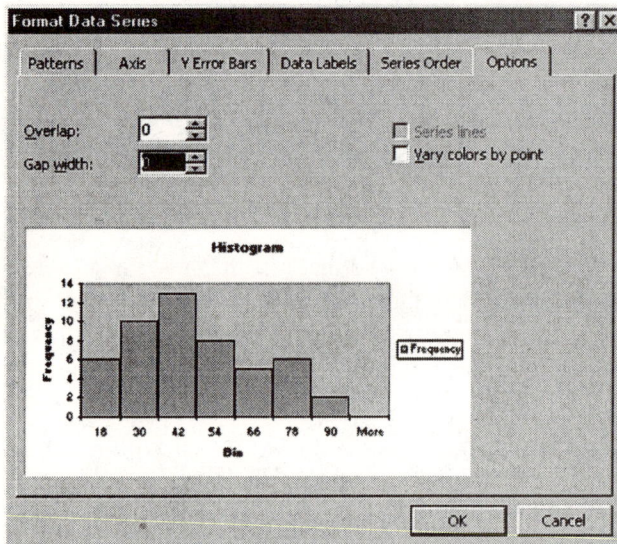

8. Change the X-axis values from upper limits to midpoints. The midpoints are displayed in a table on page 35 of your textbook. Enter these midpoints in column C of the Excel worksheet as shown below.

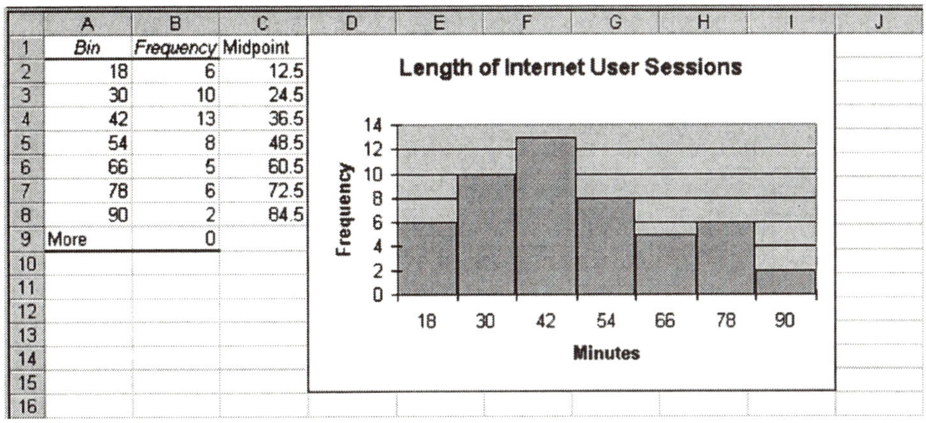

9. **Right click** on a vertical bar. Select **Source Data** from the shortcut menu that appears.

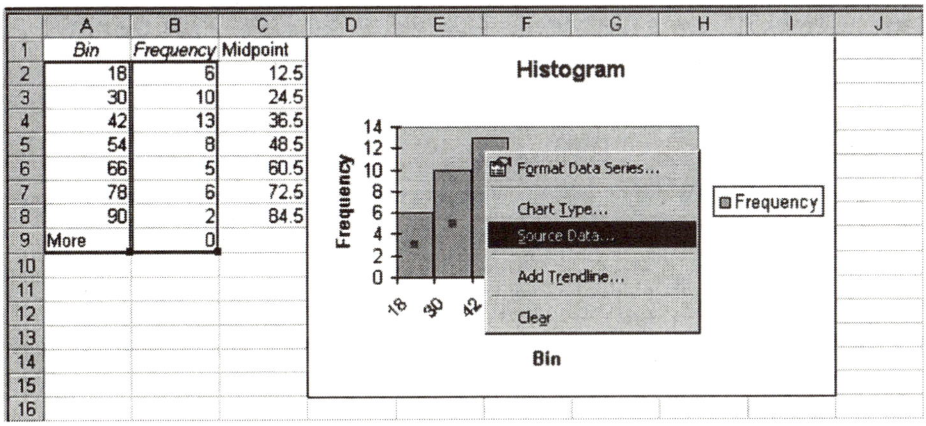

10. Click on the **Series** tab at the top of the Source Data dialog box. The ranges displayed in the Values field and the Category (X) axis labels field refer to the frequency distribution table in the top left of the worksheet. You do not want to include row 9, because that is the row containing information related to the "More" category. You also want the midpoint values in column C to be displayed on the X axis rather than the column A bin values. First, change the 9 to 8 in the **Values** field so that the entry reads **=Sheet1!$B$2:$B$8**. Next, edit the **Category (X) axis labels** field so that the entry reads **=Sheet1!$C$2:$C$8**. Click **OK**.

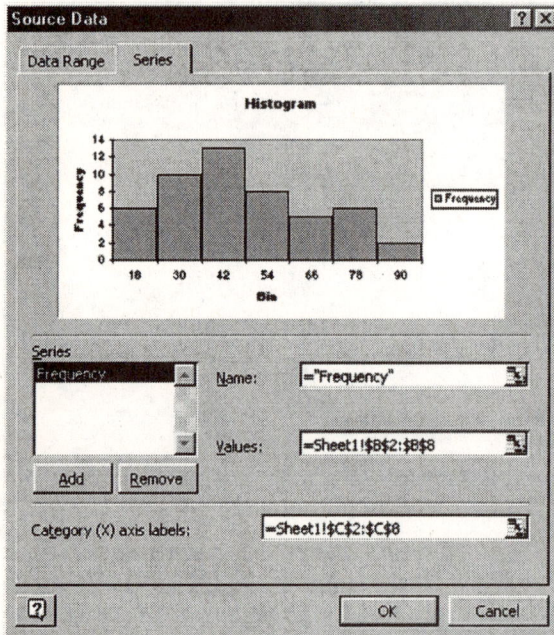

11. You will use Chart Options to modify three aspects of the histogram: Titles, gridlines, and legend. Right click in the gray plot area of the chart and select **Chart Options** from the shortcut menu that appears.

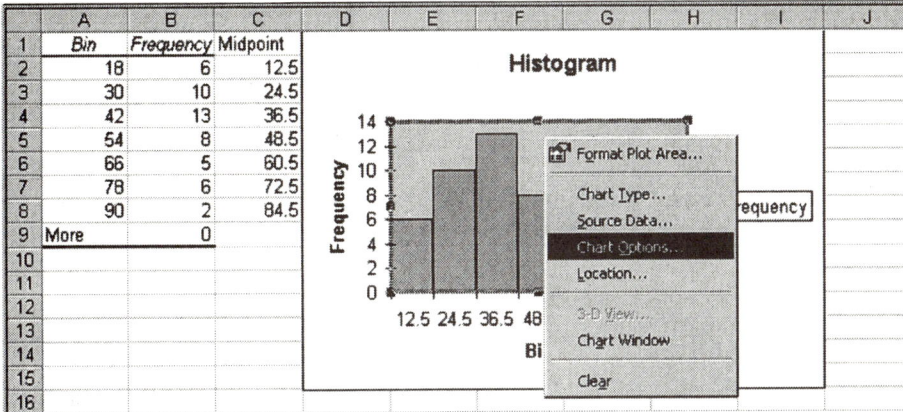

12. Click on the **Titles** tab at the top of the Chart Options dialog box. Change the Chart title from "Histogram" to **Internet Usage**. Change the Category (X) axis label from "Bin" to **Time online (in minutes)**.

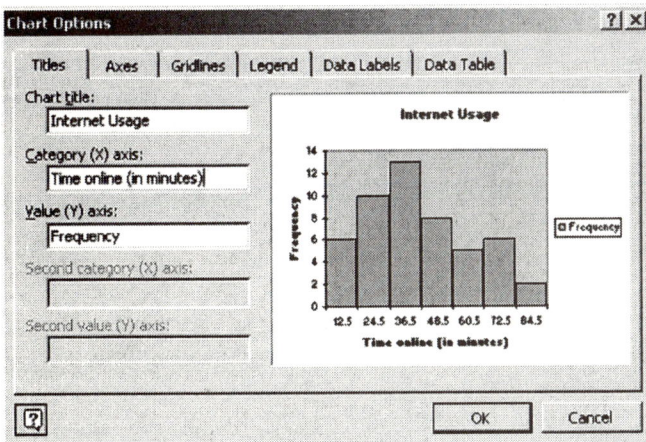

13. Click on the **Gridlines** tab at the top of the Chart Options dialog box. Under Value (Y) axis, click in the **Major gridlines** box so that a checkmark appears there.

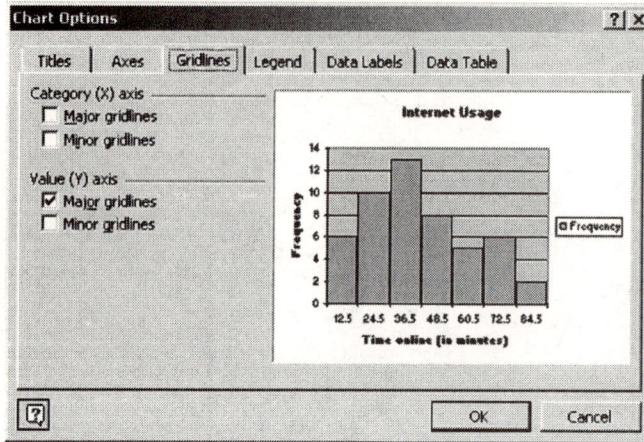

14. Click on the **Legend** tab. To remove the frequency legend displayed at the right of the histogram chart, click in the **Show legend** box to remove the checkmark. Click **OK**. Your histogram chart should now appear similar to the one displayed below.

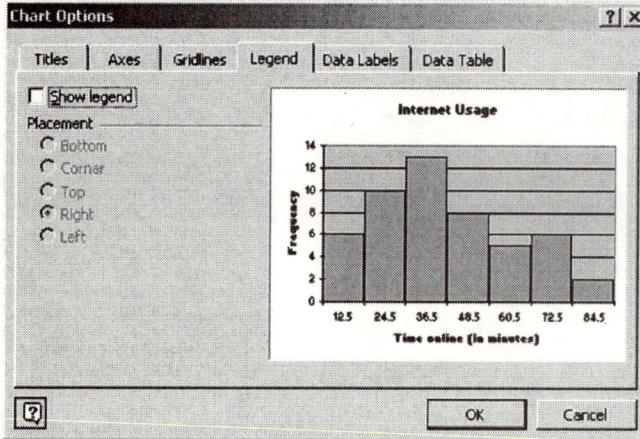

► Exercise 17 (pg. 43)	Constructing a Frequency Distribution and Frequency Histogram for the July Sales Data

1. Open worksheet "ex2_1-17" in the Chapter 2 folder.

2. Sort the data so that it is easy to identify the minimum and maximum data entries. In the sorted data set, you see that the minimum sales value is 1000 and that the maximum is 7119.

For instructions on how to sort, see Sections GS 5.1 and GS 5.2.

---

3. Calculate class width using the formula given in your textbook:

$$\text{Class width} = \frac{\text{Maximum data entry} - \text{Minimum data entry}}{\text{Number of classes}}$$

The exercise instructs you to use 6 for the number of classes.

$$\text{Class width} = \frac{7119 - 1000}{6} = 1019.83 \text{. Round up to } 1020.$$

4. The textbook instructs you to use the minimum data entry as the lower limit of the first class. To find the remaining lower limits, add the class width of 1020 to the lower limit of each previous class. Enter **Lower Limit** in cell B1 of the worksheet, and enter **1000** in cell B2. You will now use a formula to compute the remaining lower limits, and you will have Excel do these calculations for you. In cell B3, enter the formula **=B2+1020** as shown below. Press **[Enter]**.

	A	B	C	D	E	F	G
1	July Sales	Lower Limit					
2	1000	1000					
3	1030	=B2+1020					

5. Click on cell **B3** where 2020 now appears and copy the formula in cell B3 to cells B4 through B8. Because 7119 is the maximum data entry, you don't need 7120 for the histogram. However, you will be using 7120 when calculating the upper limit for the last class interval.

	A	B	C	D	E	F	G
1	July Sales	Lower Limit					
2	1000	1000					
3	1030	2020					
4	1077	3040					
5	1355	4060					
6	1500	5080					
7	1512	6100					
8	1643	7120					

6. Enter **Upper Limit** in cell C1 of the worksheet. The textbook says that the upper limit is equal to one less than the lower limit of the next higher class. You will use a formula to compute the upper limits, and you will have Excel do the computations for you. Click in cell C2 and enter the formula **=B3-1** as shown below. Press **[Enter]**.

	A	B	C	D	E	F	G
1	July Sales	Lower Limit	Upper Limit				
2	1000	1000	=B3-1				
3	1030	2020					

7. Copy the formula in cell C2 (where 2019 now appears) to cells C3 through C7. These are the upper limits that you will use as bins for constructing the histogram chart.

	A	B	C	D	E	F	G
1	July Sales	Lower Limit	Upper Limit				
2	1000	1000	2019				
3	1030	2020	3039				
4	1077	3040	4059				
5	1355	4060	5079				
6	1500	5080	6099				
7	1512	6100	7119				
8	1643	7120					

8. Click **Tools** → **Data Analysis**. Select **Histogram** and click **OK**.

If Data Analysis does not appear as a choice in the Tools menu, you will need to load the Microsoft Excel Analysis ToolPak add-in. Follow the procedure in Section GS 8.1 before continuing

---

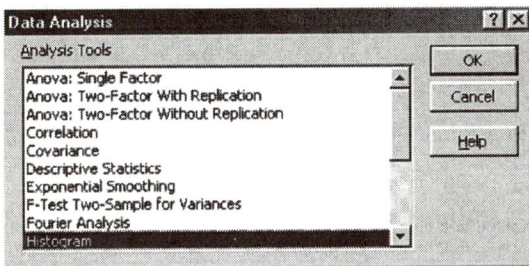

9. Complete the fields in the Histogram dialog box as shown below. The July sales data are located in cells A1 through A23 of the worksheet. The bins (upper limits) are located in cells C1 through C7. The top cells in these ranges are labels—"July Sales" and "Upper Limit," so click in the Labels box to place a checkmark there. The output will be placed in a new Excel workbook. The output will include a chart. Click **OK**.

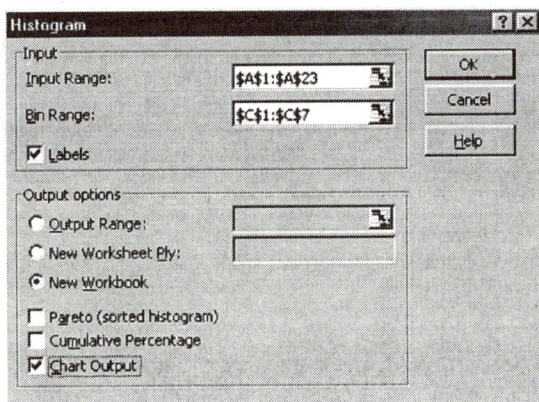

You should see output similar to the output displayed below.

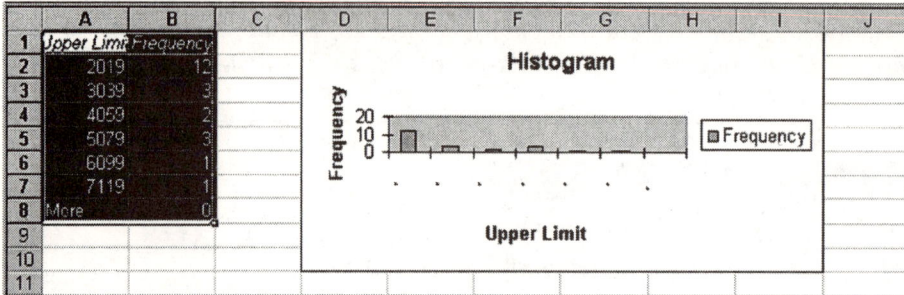

	A	B	C	D	E	F	G	H	I	J
1	Upper Limit	Frequency								
2	2019	12								
3	3039	3								
4	4059	2								
5	5079	3								
6	6099	1								
7	7119	1								
8	More	0								
9										
10										
11										

10. You will now follow steps to modify the histogram so that it is displayed in a more accurate and informative manner. First, make the chart taller so that it is easier to read. To do this, first click within the figure near a border. Black square handles appear. Click on the center handle on the bottom border of the figure and drag it down a few rows.

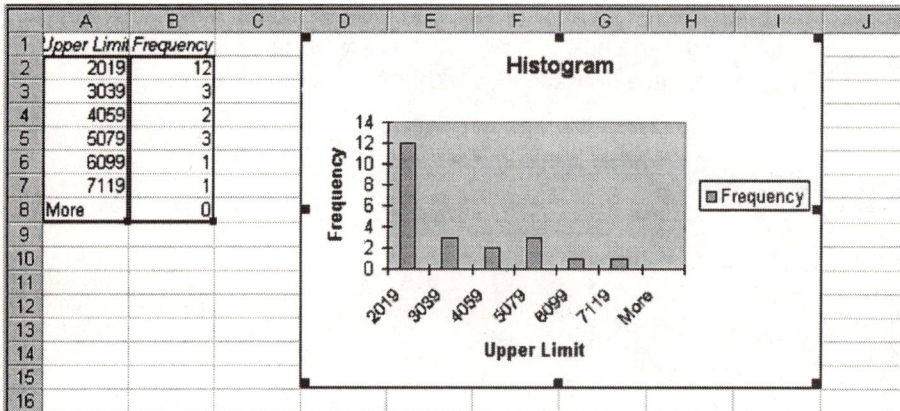

	A	B	C	D	E	F	G	H	I	J
1	Upper Limit	Frequency								
2	2019	12								
3	3039	3								
4	4059	2								
5	5079	3								
6	6099	1								
7	7119	1								
8	More	0								
9										
10										
11										
12										
13										
14										
15										
16										

11. Remove the space between the vertical bars. **Right** click on one of the vertical bars. Select **Format Data Series** from the shortcut menu that appears.

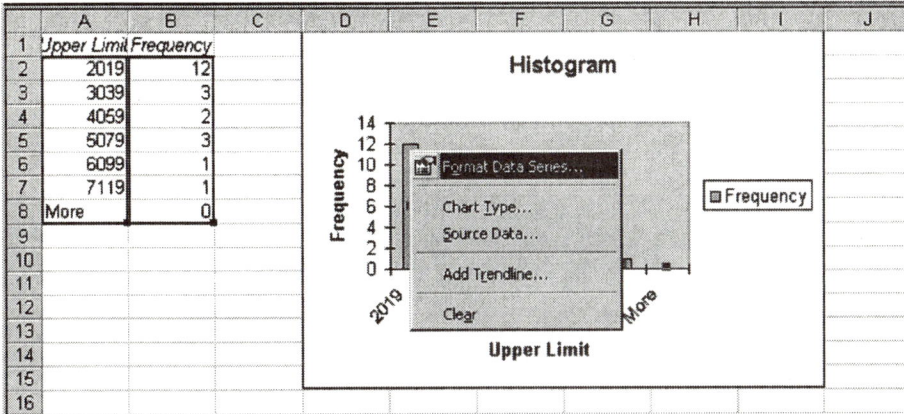

12. Click on the **Options** tab at the top of the Format Data Series dialog box. Change the value in the Gap width box to 0. Click **OK**.

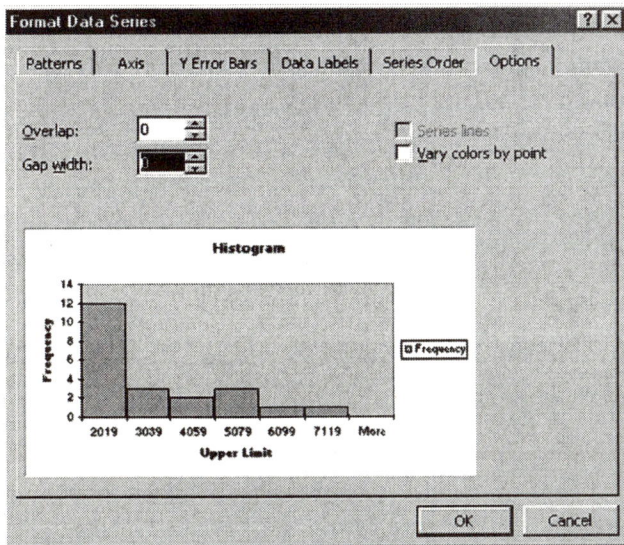

13. Delete the word "More" from the X axis. **Right click** on a vertical bar. Select **Source Data** from the shortcut menu that appears.

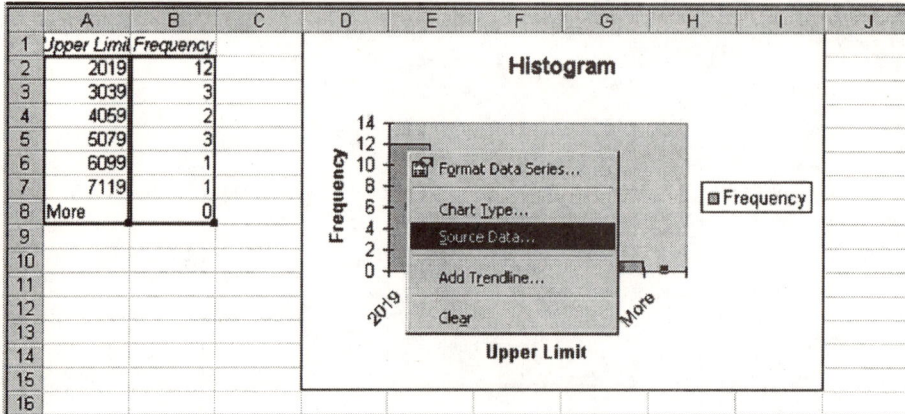

	A	B	C	D	E	F	G	H	I	J
1	Upper Limit	Frequency								
2	2019	12								
3	3039	3								
4	4059	2								
5	5079	3								
6	6099	1								
7	7119	1								
8	More	0								
9										
10										
11										
12										
13										
14										
15										
16										

14. Click on the **Series** tab at the top of the Source Data dialog box. You do not want to include row 8 of the frequency distribution because that is the row containing information related to the "More" category. Change the 8 to 7 in the **Values** field so that it reads **=Sheet1!$B$2:$B$7**. Change the 8 to 7 in the **Category (X) axis labels** field so that it reads **=Sheet1!$A$2:$A$7**. Click **OK**.

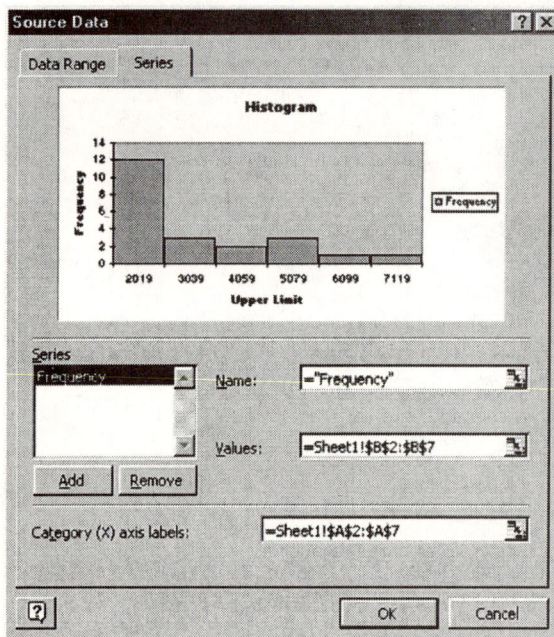

15. You will use Chart Options to modify three aspects of the histogram: Titles, gridlines, and legend. **Right click** in the gray plot area of the chart and select **Chart Options** from the shortcut menu that appears.

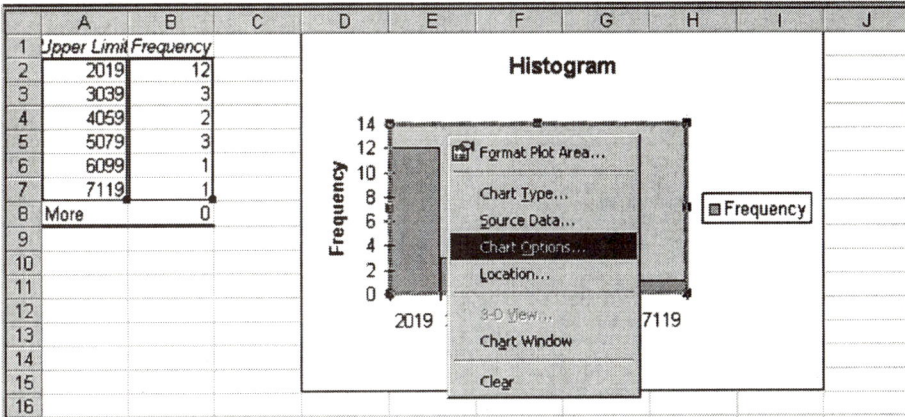

16. Click on the **Titles** tab at the top of the Chart Options dialog box. Change the Chart title from "Histogram" to **July Sales**. Change the Category (X) label from "Upper Limit" to **Dollars**.

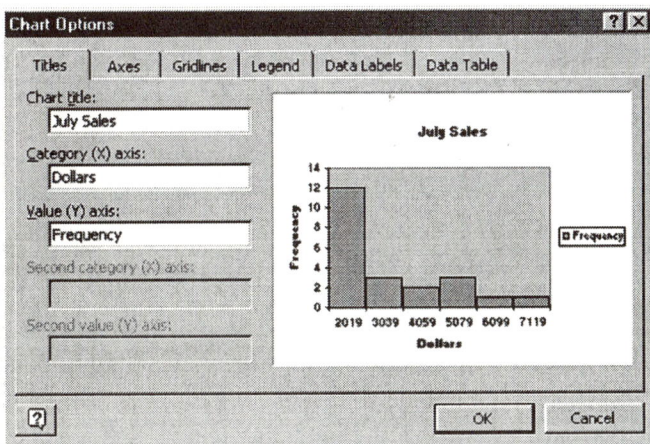

17. Click on the **Gridlines** tab at the top of the Chart Options dialog box. Under Value (Y) axis, click in the **Major gridlines** box.

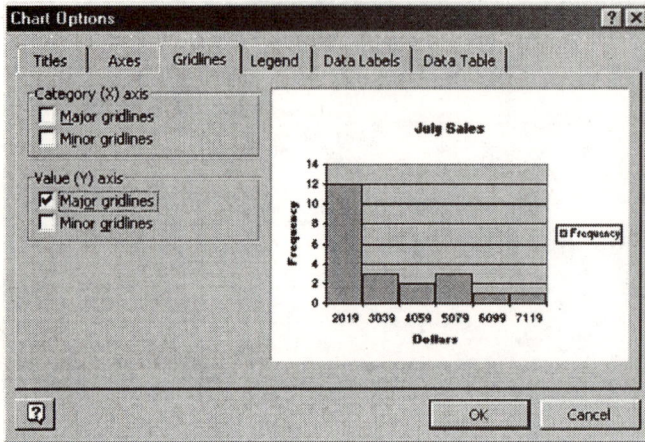

18. Click on the **Legend** tab. To remove the frequency legend displayed at the right of the histogram chart, click in the **Show legend** box to remove the checkmark. Click **OK**. Your histogram chart should now appear similar to the one displayed below.

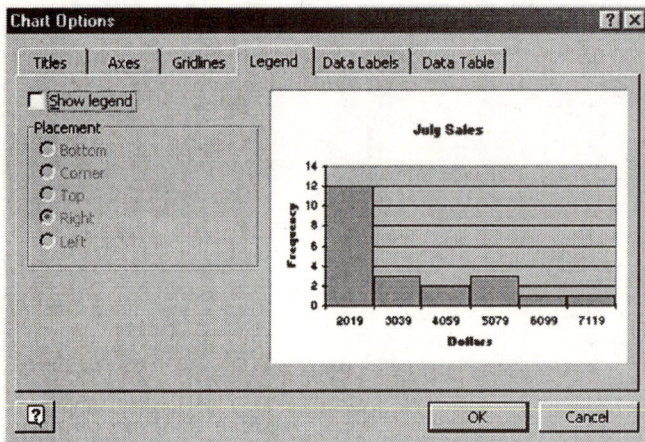

Section 2.2

▶ Example 1 (pg. 46) | Constructing a Stem-and-Leaf Plot

If the PHStat2 add-in has not been loaded, you will need to load it before continuing. Follow the instructions in Section GS 8.2.

1. Open worksheet "AL_RBIs" in the Chapter 2 folder.

2. At the top of the screen, click **PHStat→Descriptive Statistics→Stem-and-Leaf Display**.

3. Complete the entries in the Stem-and-Leaf Display dialog box as shown below. Click **OK**.

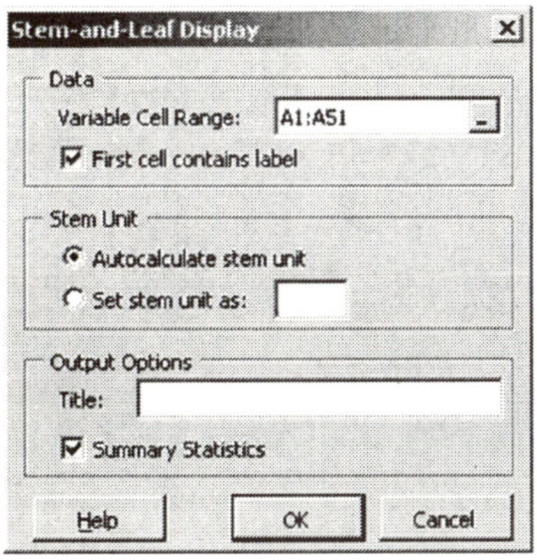

The output is displayed in a new sheet named "StemLeafPlot."

	A	B	C	D	E	F	G
1				Stem-and-Leaf Display			
2				for A.L. RBIs			
3				Stem unit: 10			
4							
5	Statistics			7	8		
6	Sample Size	50		8			
7	Mean	125.14		9			
8	Median	123.5		10	5 8 9 9 9		
9	Std. Deviation	14.53919		11	2 2 2 3 4 6 7 8 8 8 9 9 9		
10	Minimum	78		12	1 1 2 2 2 3 4 4 6 6 6 6 6 9 9		
11	Maximum	159		13	0 2 3 3 4 9 9		
12				14	0 2 4 5 5 7 8		
13				15	5 9		
14							

◀

▶ **Example 4   (pg. 49)**   **Constructing a Pie Chart**
_____

1. Open a new, blank Excel worksheet and enter the motor vehicle frequency table as shown below.

	A	B	C	D	E	F	G
1	Vehicle type	Killed					
2	Cars	25,063					
3	Trucks	9,409					
4	Motorcycles	3,141					
5	Other	474					

2. Click on any cell within the table. Then click **Insert → Chart**.

_____

If a cell within the table is activated when you select Chart, the data range will automatically be entered into the Chart dialog box.

_____

3. Under Chart type, in the Chart Type dialog box, select **Pie**. Under Chart sub-type, select the first one in the top row by clicking on it. Click **Next** at the bottom of the dialog box.

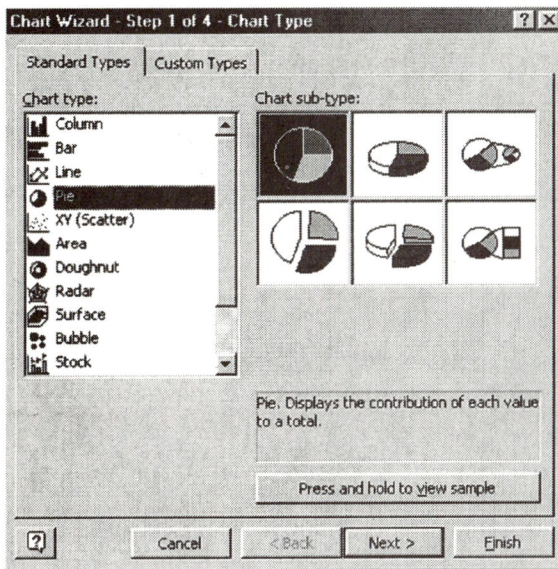

4. Check the accuracy of the data range in the Chart Source Data dialog box. It should read **=Sheet1!$A$1:$B$5**. Make any necessary corrections. Then click **Next**.

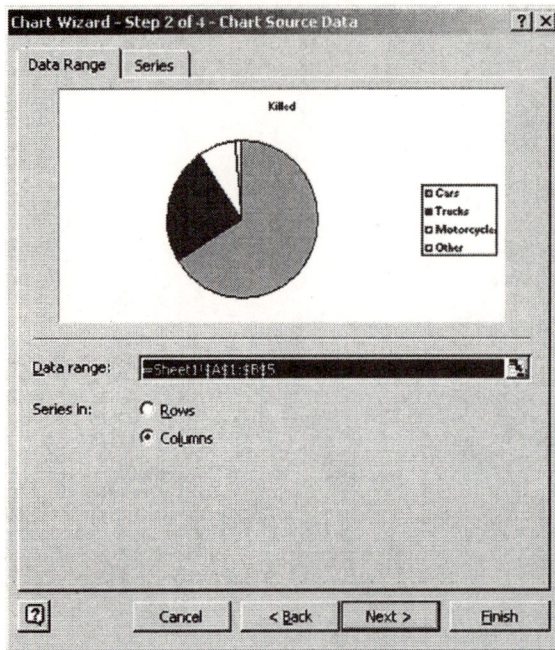

5. Click the **Titles** tab at the top of the Chart Options dialog box. Change the title from "Killed" to **Motor Vehicle Occupants Killed**.

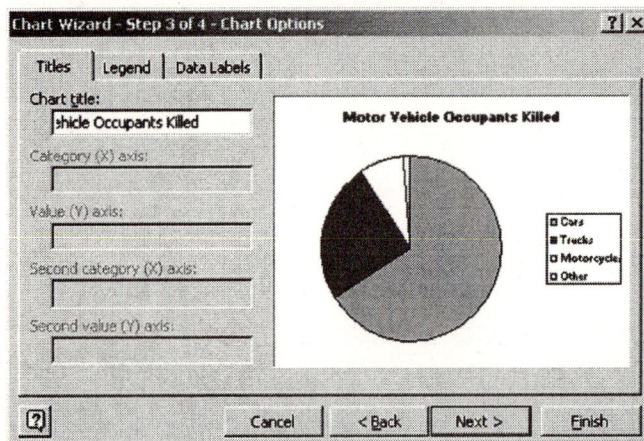

6. Click the **Data Labels** tab at the top of the Chart Options dialog box. Click the button next to **Show label and percent** so that a black dot appears there. Click **Next**.

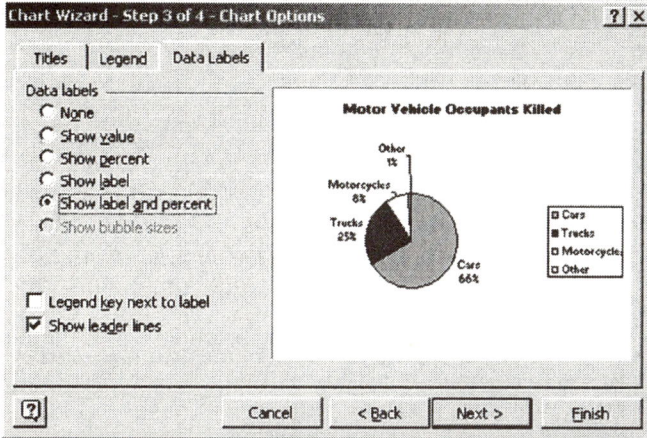

7. The Chart Location dialog box presents two options for placing your chart output. For this example, select **As object in**. Click **Finish**.

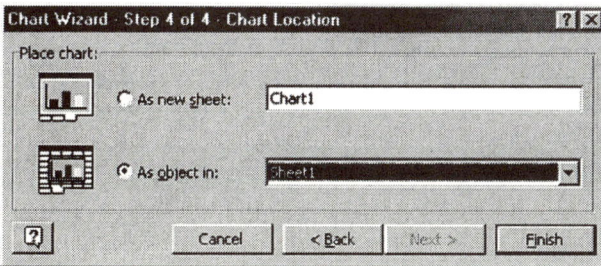

8. In the chart shown below, the final letter of "Motorcycles" is shown below the rest of the word.

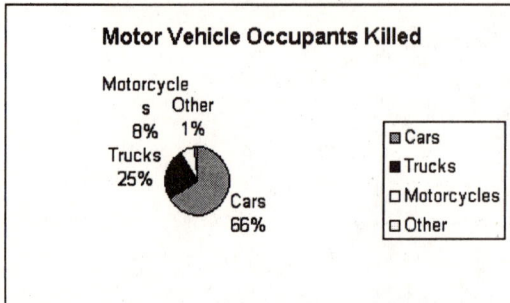

To fix this, first **right click** directly on the Motorcycle label. Then click on **Format Data Labels** in the shortcut menu that appears.

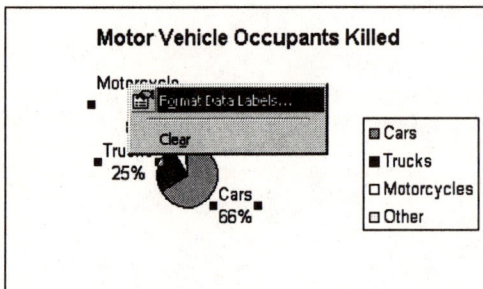

9. Click the **Font** tab at the top of the Format Data Labels dialog box. In the Size: window, change the size from 10 to **9**. Click **OK**.

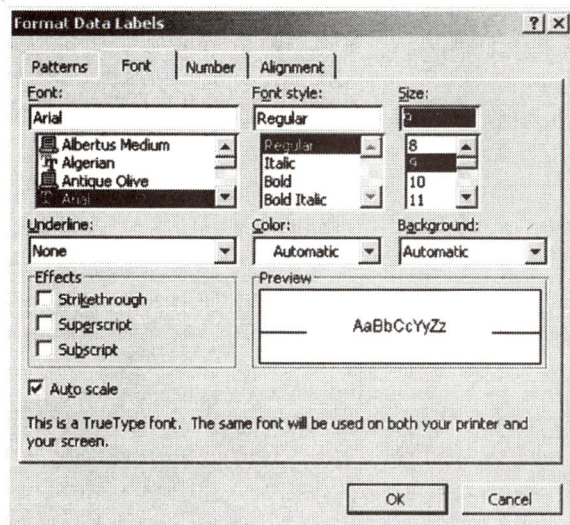

10. The chart title is now on top of a data label.   To provide more space at the top of the chart, first click within the figure near the top border. Black square handles appear. Click on the center handle on the top border of the figure and drag it up a few rows.

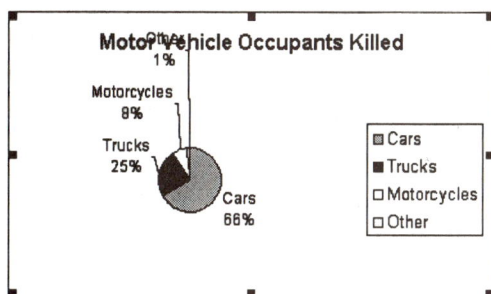

Your pie chart for the motor vehicle data should look similar to the one shown below.

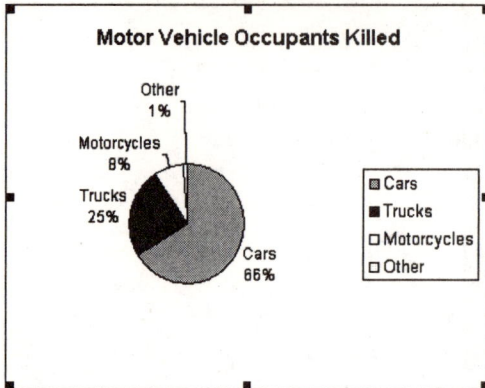

Motor Vehicle Occupants Killed

Other 1%
Motorcycles 8%
Trucks 25%
Cars 66%

- Cars
- Trucks
- Motorcycles
- Other

◄

► Example 5 (pg. 50)	Constructing a Pareto Chart

1. Open a new, blank Excel worksheet, and enter the inventory shrinkage information as shown below.

	A	B	C	D	E	F	G
1	Cause	Millions of dollars					
2	Administrative error	7.8					
3	Employee theft	15.6					
4	Shoplifting	14.7					
5	Vendor fraud	2.9					

2. In a Pareto chart, the vertical bars are placed in order of decreasing height. Excel's Chart function will display the information in the order in which it appears in the worksheet. So, you need to sort the information in descending order by Millions of dollars.

For instructions on how to sort, see Sections GS 5.1 and GS 5.2.

	A	B	C	D	E	F	G
1	Cause	Millions of dollars					
2	Employee theft	15.6					
3	Shoplifting	14.7					
4	Administrative error	7.8					
5	Vendor fraud	2.9					

3. After sorting the data, click on any cell within the data table. Then click **Insert →
   Chart**.

4. In the Chart Type dialog box, select a **Column** chart type. Click on the top left
   chart sub-type to select it. Click **Next**.

5. Check the range displayed in the Chart Source Data dialog box to make sure it is
   accurate. It should read =Sheet1!$A$1:$B$5. Make any necessary corrections.
   Then click **Next**.

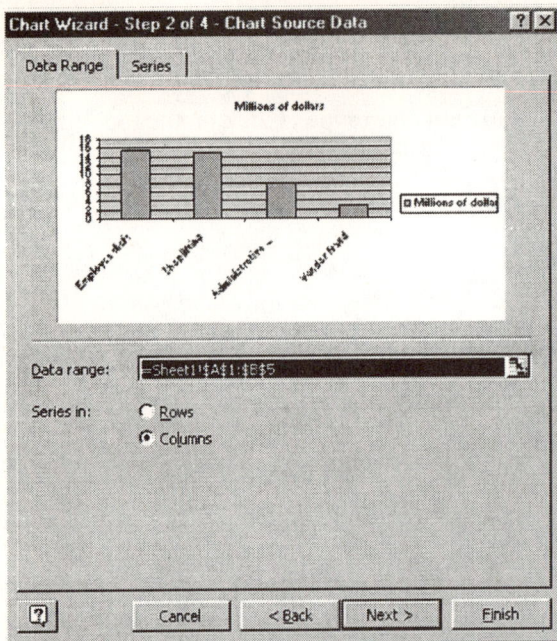

6. Click the **Titles** tab at the top of the Chart Options dialog box. In the Chart title field, enter **Causes of Inventory Shrinkage**. In the Category (X) axis field, enter **Cause**. In the Value (Y) axis field, enter **Millions of dollars**.

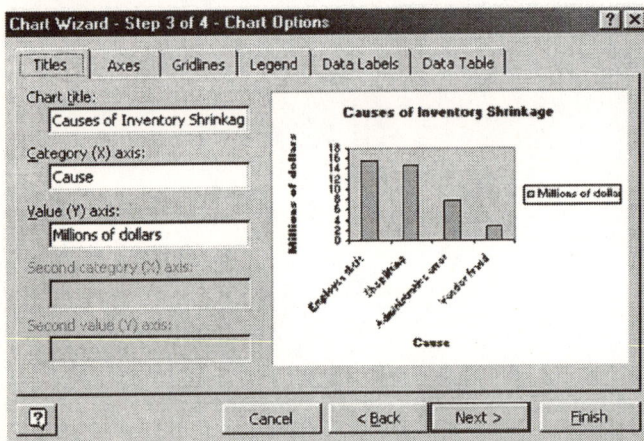

7. Click the **Gridlines** tab at the top of the Chart Options dialog box. Below Value (Y) axis, click in the box to the left of **Major gridlines** so that a checkmark appears there.

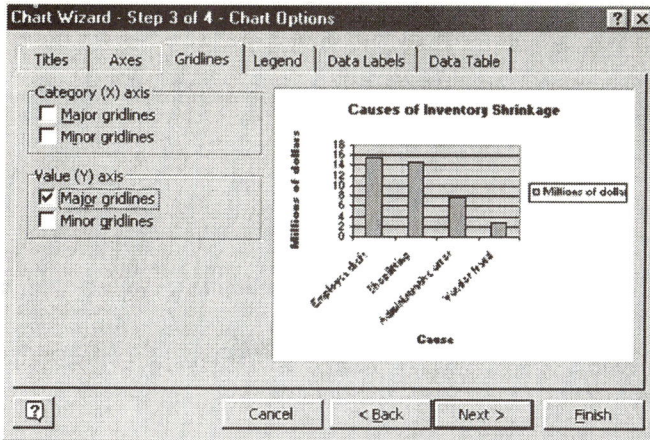

8. Click the **Legend** tab at the top of the Chart Options dialog box. Click in the box to the left of **Show legend** so that no checkmark appears there. This will remove the Millions of dollars legend on the right side of the chart. Click **Next**.

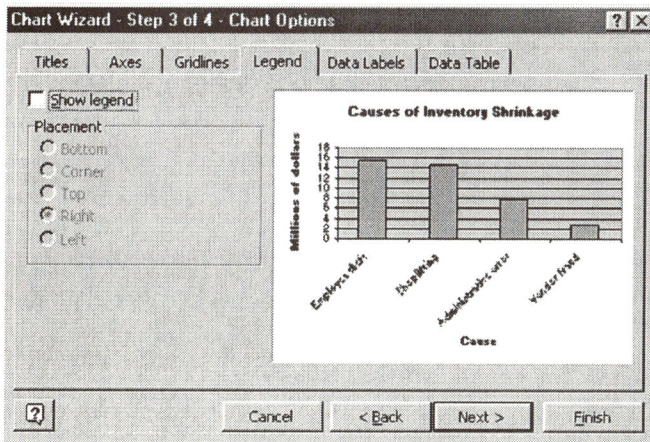

9. The Chart Location dialog box presents two options for placement of the chart. For this example, select **As object in**. Click **Finish**.

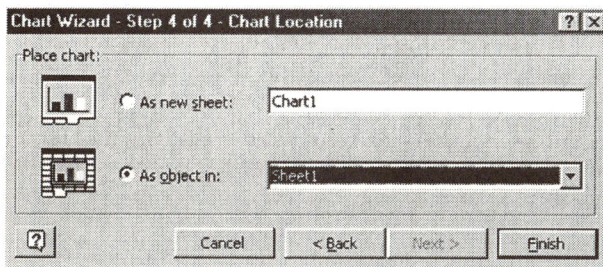

10. Make the chart taller so that it is easier to read. To do this, first click within the figure near a border so that black square handles appear. Click on the center handle at the bottom border of the figure and drag it down a few rows. Your chart should look similar to the one shown below.

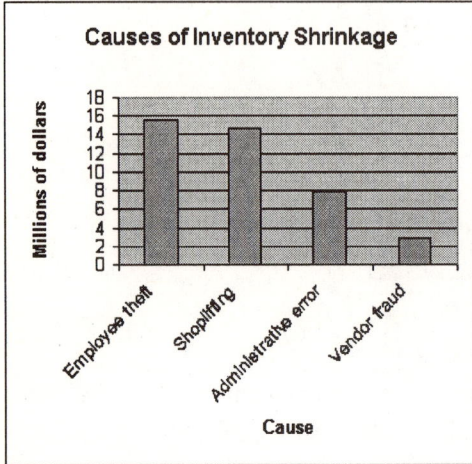

11. Remove the space between the vertical bars. **Right click** on one of the vertical bars. Select **Format Data Series** from the shortcut menu that appears.

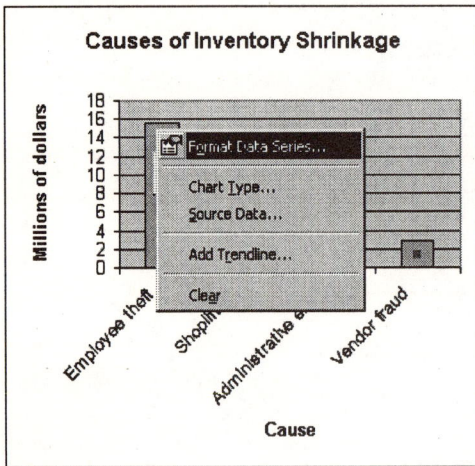

12. Click the **Options** tab at the top of the Format Data Series dialog box. Change the value in the Gap width box to 0. Click **OK**. Your histogram chart should now look similar to the one displayed below.

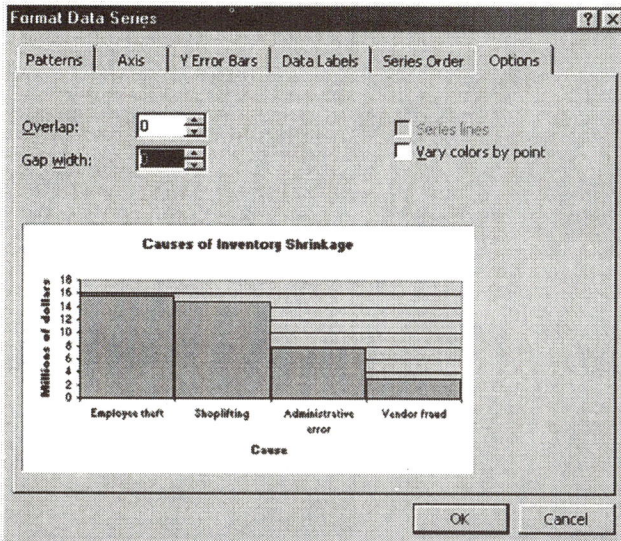

▶ Example 7 (pg. 52) | Constructing a Time Series Chart

1. Open worksheet "cellphone" in the Chapter 2 folder.

2. Click on any cell within the table. At the top of the screen, click **Insert** and select **Chart** from the menu that appears.

3.   In the Chart Type dialog box, select the **XY (Scatter)** chart type by clicking on it. Then click on the topmost chart sub-type. Click **Next**.

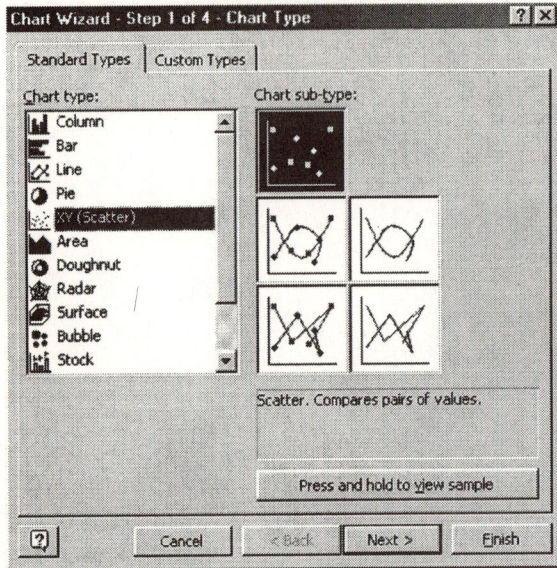

4. In the Chart Source Data dialog box, edit the data range so that it does not include the average bill data in column C of the worksheet. The entry in the data range field should be **=cellphone!$A$1:$B$14**. Click **Next**.

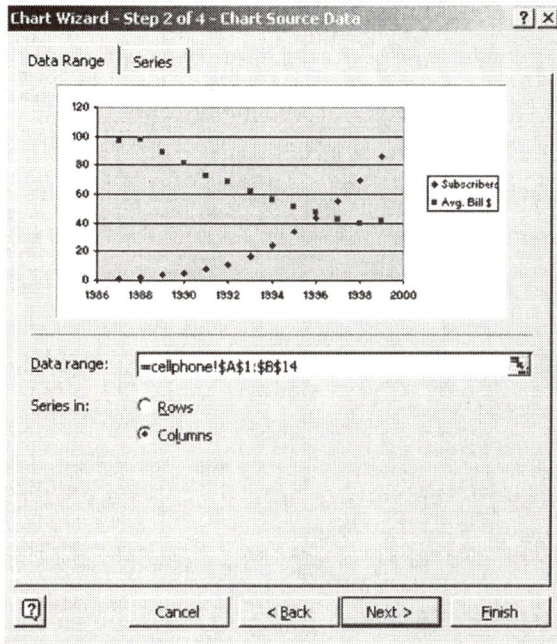

5. Click the **Titles** tab at the top of the Chart Options dialog box. In the Chart title field, change the title to **Cellular Telephone Subscribers by Year**. In the Value (X) Axis field, enter **Year**. In the Value (Y) axis field, enter **Subscribers (in millions)**.

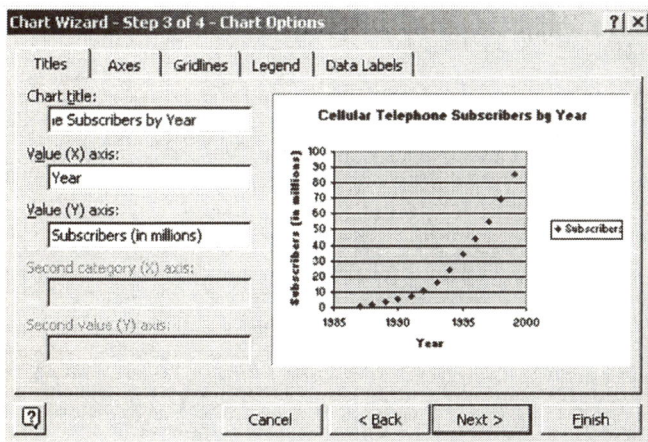

6. Click the **Legend** tab at the top of the Chart Options dialog box. Click in the box to the left of **Show Legend** so that a checkmark does not appear there. This removes the Subscribers legend from the right side of the scatter plot. Click **Next**.

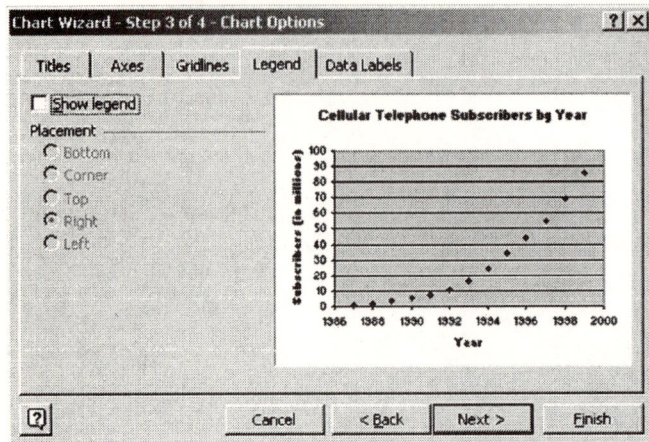

7. The Chart Location dialog box presents two options for placing the scatter plot. For this example, select **As object in** the same worksheet as the data. To do this, click the button to the left of **As object in** so that a black dot appears there. Click **Finish**.

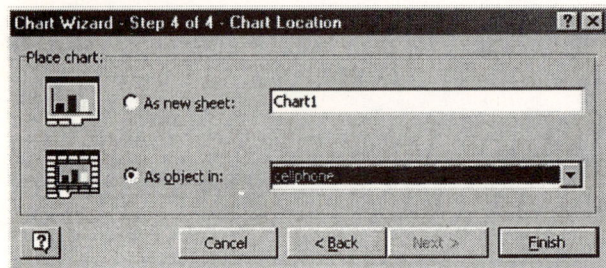

8. The scatter plot that you will obtain is quite compressed. Make it taller so that it is easier to read. To do this, first click within the figure near a border. Black square handles appear. Then click on the center handle on the bottom border of the figure and drag it down a few rows.

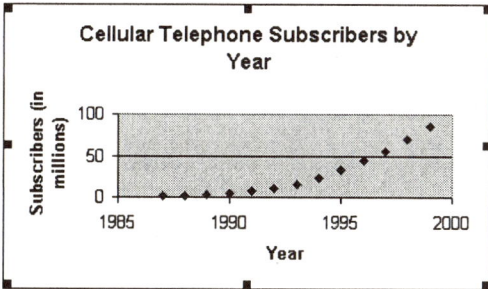

Your scatter plot should now look similar to the one shown below.

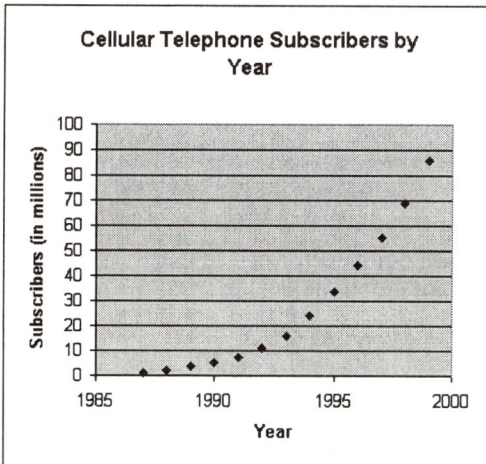

Section 2.3

▶ Example 6 (pg. 60)	Finding the Mean, Median, and Mode

If the PHStat2 add-in has not been loaded, you will need to load it before continuing. Follow the instructions in Section GS 8.2.

1. Open worksheet "ages" in the Chapter 2 folder.

2. At the top of the screen, click **Tools → Data Analysis**. Select **Descriptive Statistics** and click **OK**.

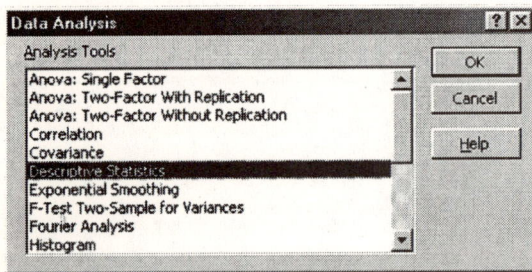

3. Complete the Descriptive Statistics dialog box as shown below. The entry in the
    Input Range field presents the worksheet location of the age data. The checkmark
    to the left of **Labels in First Row** lets Excel know that the entry in cell A1 is a label
    and is not to be included in the computations. The output will be placed in a **New
    Worksheet**. The checkmark to the left of **Summary statistics** requests that the
    output include summary statistics for the specified set of data. Click **OK**.

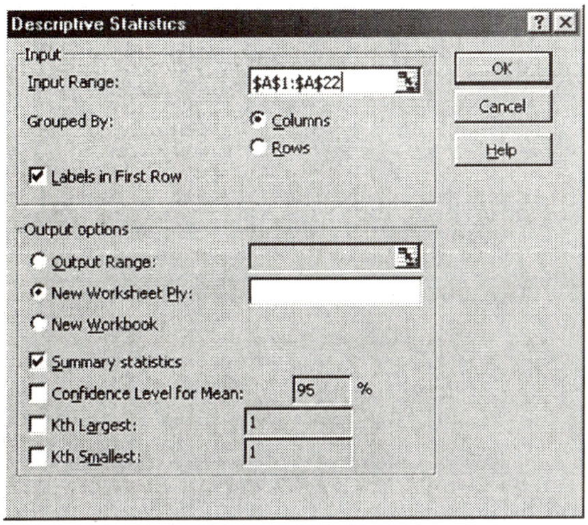

4. The output is placed in a worksheet given the default name "Sheet1." You will want to widen column A of the output so that you can read the labels. Your output should be similar to the output shown below. The mean of the sample is 23.75, the median is 21.5, and the mode is 20.

Be careful when using the value of the mode reported in the Descriptive Statistics output. If there is a tie for the mode, Excel reports only the first value that occurs in the data set. Therefore, it is always a good idea to construct a frequency distribution table to go along with the Descriptive Statistics output.

	A	B
1	Ages of students in Class	
2		
3	Mean	23.75
4	Standard Error	2.193141
5	Median	21.5
6	Mode	20
7	Standard Deviation	9.808026
8	Sample Variance	96.19737
9	Kurtosis	19.0739
10	Skewness	4.324551
11	Range	45
12	Minimum	20
13	Maximum	65
14	Sum	475
15	Count	20

5. Construct a frequency distribution table to see if the data set is unimodal or multimodal. Go back to the worksheet containing the age data. To do this, click on the **ages** sheet tab near the bottom of the screen. Then select **PHStat** → **Descriptive Statistics** → **One-Way Tables & Charts**.

6. Complete the One-Way Tables & Charts dialog box as shown below. Click **OK**.

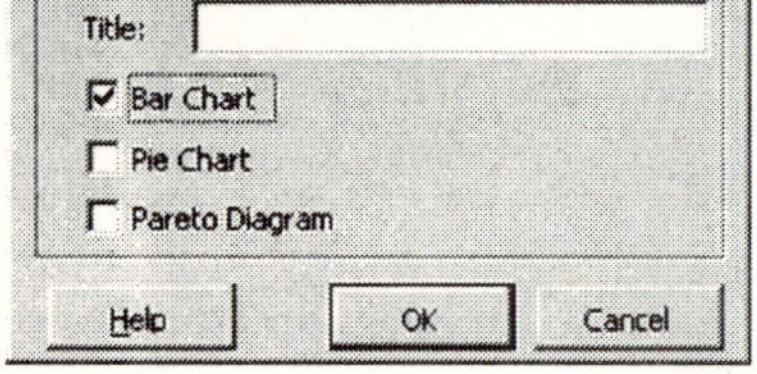

One-Way Tables & Charts

Data
Type of Data:
- ⊙ Raw Categorical Data
- ○ Table of Frequencies

Raw Data Cell Range: A1:A21

☑ First cell contains label

Output Options
Title:
☑ Bar Chart
☐ Pie Chart
☐ Pareto Diagram

Help   OK   Cancel

The bar chart is displayed in a worksheet named "Bar Chart."

Bar Chart

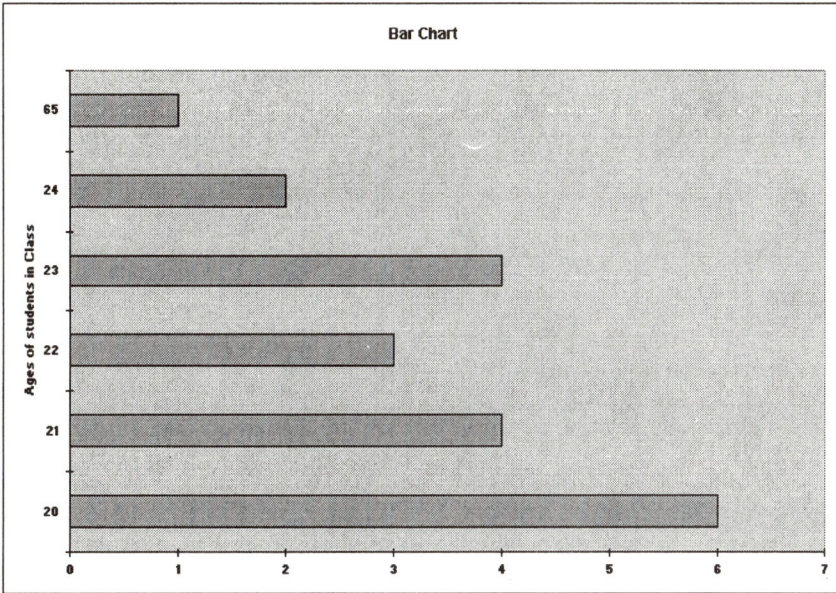

7. Click on the **OneWay Table** sheet tab at the bottom of the screen. This sheet contains a frequency distribution of the ages. You can see by looking at the bar chart and the frequency distribution that the age distribution is unimodal. The modal value of 20 has a frequency of six.

	A	B	C	D	E	F	G
1	One-Way Summary Table						
2							
3	Count of Ages of students in Class						
4	Ages of students in Class	Total					
5	20	6					
6	21	4					
7	22	3					
8	23	4					
9	24	2					
10	65	1					
11	Grand Total	20					

Section 2.4

▶ **Example 5** (pg. 74)	Finding the Standard Deviation

1. Open worksheet rentrate in the Chapter 2 folder.

2. Click **Tools** → **Data Analysis**. Select **Descriptive Statistics** and click **OK**.

If Data Analysis does not appear as a choice in the Tools menu, you will need to load the Microsoft Excel Analysis ToolPak add-in. Follow the procedure in Section GS 8.1 before continuing.

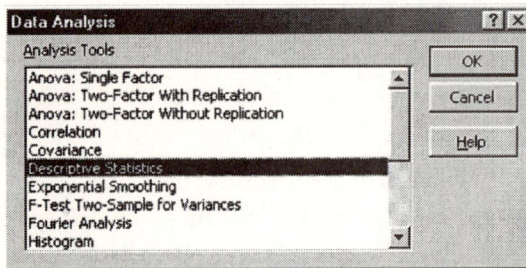

3. Complete the Descriptive Statistics dialog box as shown below. Click **OK**.

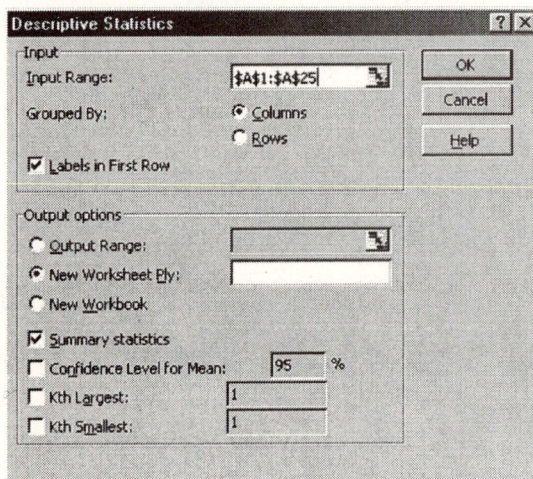

You will want to make Column A of the output wider so that you can read the labels. Your Descriptive Statistics output should be similar to the output shown below. The mean is 33.73 and the standard deviation is 5.09.

	A	B
1	*Miami rates*	
2		
3	Mean	33.72917
4	Standard Error	1.038864
5	Median	35.375
6	Mode	37
7	Standard Deviation	5.089373
8	Sample Variance	25.90172
9	Kurtosis	-0.74282
10	Skewness	-0.70345
11	Range	16.75
12	Minimum	23.75
13	Maximum	40.5
14	Sum	809.5
15	Count	24

◀

Section 2.5

▶ Example 2 (pg. 88)	Finding Quartiles

1. Open worksheet "tuition" in the Chapter 2 folder. Key in labels for the quartiles as shown below. Then click in cell **C2** to place the first quartile there.

	A	B	C	D	E	F	G
1	tuition		Quartile 1				
2	23						
3	25						
4	30		Quartile 2				
5	23						
6	20						
7	22		Quartile 3				

2. You will be using the QUARTILE function to obtain the first, second, and third quartiles for the tuition data. At the top of the screen, click **Insert → Function**.

3.   Under Function category, select **Statistical** by clicking on it. Under Function name, select **QUARTILE**. Click **OK**.

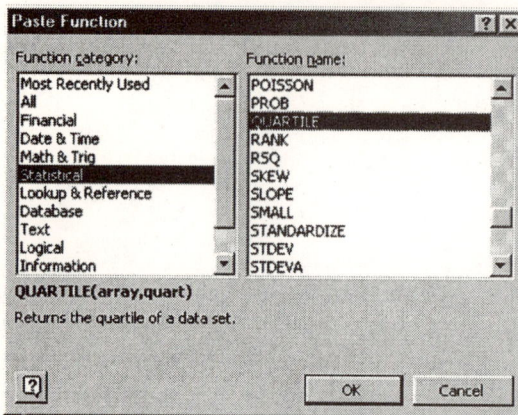

4.   Complete the QUARTILE dialog box as shown below. Click **OK**.

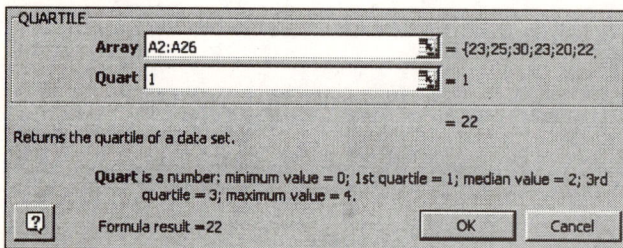

5.   Click in cell **C5** to place the second quartile there.

6.   At the top of the screen, click **Insert → Function**.

7.   Under Function category, select **Statistical** by clicking on it. Under Function name, select **QUARTILE**. Click **OK**.

8. Complete the QUARTILE dialog box as shown below. Click **OK**.

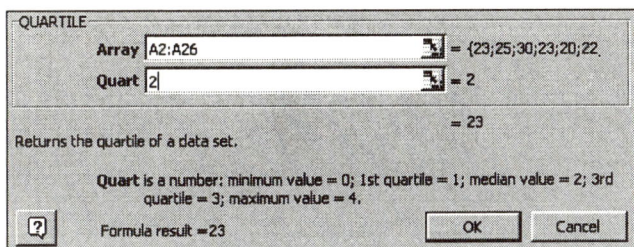

```
QUARTILE
        Array  A2:A26                        ▼ = {23;25;30;23;20;22
        Quart  2                             ▼ = 2

                                       = 23
Returns the quartile of a data set.

        Quart is a number: minimum value = 0; 1st quartile = 1; median value = 2; 3rd
                quartile = 3; maximum value = 4.
   ?    Formula result =23              [ OK ]    [ Cancel ]
```

9. Click in cell **C8** to place the third quartile there.

10. At the top of the screen, click **Insert → Function**.

11. Under Function category, select **Statistical** by clicking on it. Under Function name, select **QUARTILE**. Click **OK**.

12. Complete the QUARTILE dialog box as shown below. Click **OK**.

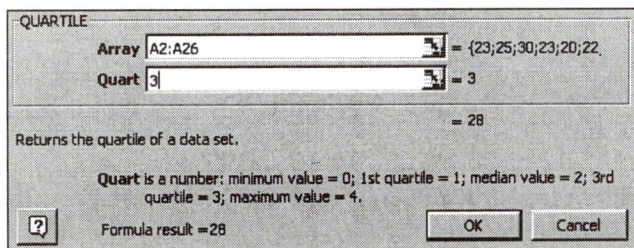

```
QUARTILE
        Array  A2:A26                        ▼ = {23;25;30;23;20;22
        Quart  3                             ▼ = 3

                                       = 28
Returns the quartile of a data set.

        Quart is a number: minimum value = 0; 1st quartile = 1; median value = 2; 3rd
                quartile = 3; maximum value = 4.
   ?    Formula result =28              [ OK ]    [ Cancel ]
```

Your output should look similar to the output displayed below.

	A	B	C	D	E	F	G
1	tuition		Quartile 1				
2	23		22				
3	25						
4	30		Quartile 2				
5	23		23				
6	20						
7	22		Quartile 3				
8	21		28				

◄

| ► Example 4 (pg. 90) | Drawing a Box-and-Whisker Plot |

If the PHStat2 add-in has not been loaded, you will need to load it before continuing. Follow the instructions in Section GS 8.2.

1. Open a new Excel worksheet and enter the test scores of 15 employees enrolled in a CPR training course as shown below.

	A	B	C	D	E	F	G
1	Test score						
2	13						
3	9						
4	18						
5	15						
6	14						
7	21						
8	7						
9	10						
10	11						
11	20						
12	5						
13	18						
14	37						
15	16						
16	17						

2. At the top of the screen, select **PHStat → Descriptive Statistics → Box-and-Whisker Plot**.

3. Complete the Box-and-Whisker Plot dialog box as shown below. Click **OK**.

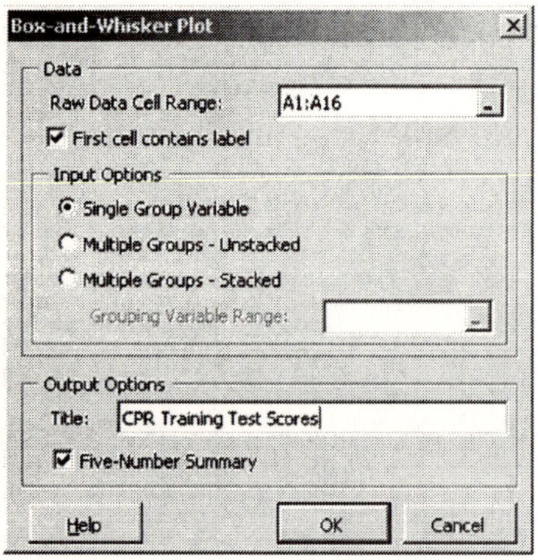

Box-and-Whisker Plot

Data
Raw Data Cell Range: A1:A16
☑ First cell contains label

Input Options
● Single Group Variable
○ Multiple Groups - Unstacked
○ Multiple Groups - Stacked
Grouping Variable Range:

Output Options
Title: CPR Training Test Scores
☑ Five-Number Summary

Help   OK   Cancel

The Five-Number Summary is displayed in a worksheet named "FiveNumbers."

	A	B	C
1	CPR Training Test Scores		
2			
3	Five-number Summary		
4	Minimum	5	
5	First Quartile	10.5	
6	Median	15	
7	Third Quartile	18	
8	Maximum	37	

The box-and-whisker plot is displayed in a worksheet named "BoxWhiskerPlot."

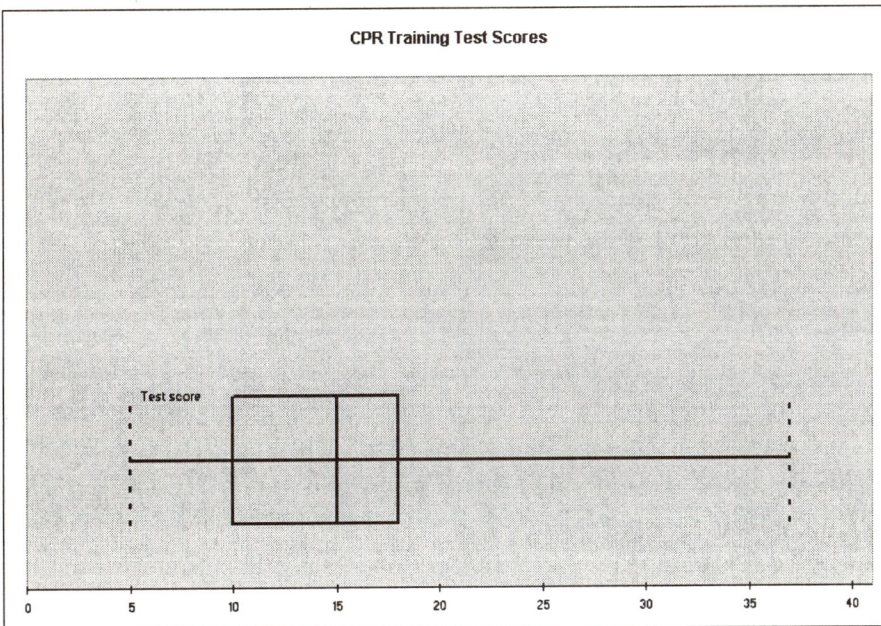

CPR Training Test Scores

Technology Lab

▶ Exercises 1 & 2 (pg. 97)	Finding the Sample Mean and the Sample Standard Deviation

1. Open worksheet "Tech2" in the Chapter 2 folder.

2. Click **Tools → Data Analysis**. Select **Descriptive Statistics** and click OK.

If Data Analysis does not appear as a choice in the Tools menu, you will need to load the Microsoft Excel Analysis ToolPak add-in. Follow the procedure in Section GS 8.1 before continuing.

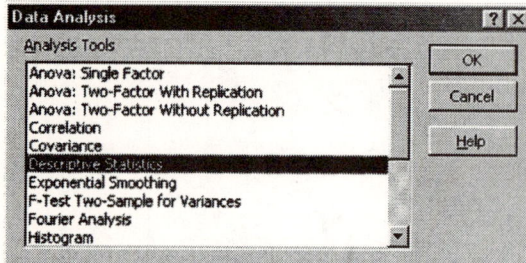

3. Complete the Descriptive Statistics dialog box as shown below. Click **OK**.

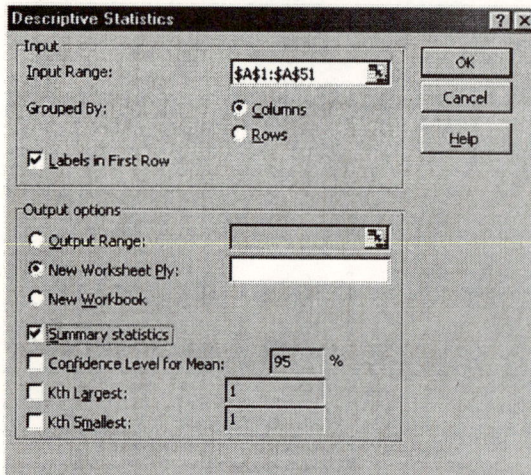

You will want to adjust the width of Column A of the output so that you can read the labels. Your Descriptive Statistics output should be similar to the output shown below. The sample mean is 2270.54 and the sample standard deviation is 653.1822.

	A	B
1	*Monthly Milk Production*	
2		
3	Mean	2270.54
4	Standard Error	92.37391
5	Median	2207
6	Mode	2207
7	Standard Deviation	653.1822
8	Sample Variance	426647
9	Kurtosis	0.567664
10	Skewness	0.549267
11	Range	3138
12	Minimum	1147
13	Maximum	4285
14	Sum	113527
15	Count	50

◀

▶ Exercises 3 & 4 (pg. 97)	Constructing a Frequency Distribution and a Frequency Histogram

1. Open worksheet "Tech2" in the Chapter 2 folder.

2. Sort the data in ascending order. In the sorted data set you can see that the minimum production is 1147 pounds. Scroll down to find the maximum production. The maximum production, displayed in cell A51, is 4285 pounds.

For instructions on how to sort, refer to Sections GS 5.1 and GS 5.2.

	A	B	C	D	E	F	G
1	Monthly Milk Production						
2	1147						
3	1230						
4	1258						

3. Enter **Lower Limit** in cell D1 of the worksheet. The textbook tells you to use the minimum value as the lower limit of the first class. Enter **1147** in cell D2. You will calculate the remaining lower limits by adding the class width of 500 to the lower limit of each previous class. You will use a formula to do these computations in the Excel worksheet. Click in cell **D3** and key in **=D2+500** as shown below. Press [**Enter**].

	A	B	C	D	E	F	G
1	Monthly Milk Production			Lower Limit			
2	1147			1147			
3	1230			=D2+500			

4. Click on cell D3 (where 1647 now appears) and copy the contents of cell D3 to cells D4 through D9. Because the maximum milk production is 4285 pounds, you have calculated one more lower limit than is needed for the histogram. The value of 4647, however, will be used when calculating the upper limit of the last class.

	A	B	C	D	E	F	G
1	Monthly Milk Production			Lower Limit			
2	1147			1147			
3	1230			1647			
4	1258			2147			
5	1294			2647			
6	1319			3147			
7	1449			3647			
8	1619			4147			
9	1647			4647			

5. Enter **Upper Limit** in cell E1. The upper limit is equal to one less than the lower limit of the next higher class. To do these calculations, you will enter a formula in the Excel worksheet. Click in cell E2 and enter the formula **=D3-1** as shown in the worksheet below. Press [**Enter**].

	A	B	C	D	E	F	G
1	Monthly Milk Production			Lower Limit	Upper Limit		
2	1147			1147	=D3-1		

6. Copy the formula in E2 (where 1646 now appears) to cells E3 through E8. You will use these upper limits for the bins when you construct the histogram chart.

	A	B	C	D	E	F	G
1	Monthly Milk Production			Lower Limit	Upper Limit		
2	1147			1147	1646		
3	1230			1647	2146		
4	1258			2147	2646		
5	1294			2647	3146		
6	1319			3147	3646		
7	1449			3647	4146		
8	1619			4147	4646		
9	1647			4647			

7. At the top of the screen, click **Tools → Data Analysis**. Select **Histogram** and click **OK**.

If Data Analysis does not appear as a choice in the Tools menu, you will need to load the Microsoft Excel Analysis ToolPak add-in. Follow the procedure in Section GS 8.1 before continuing.

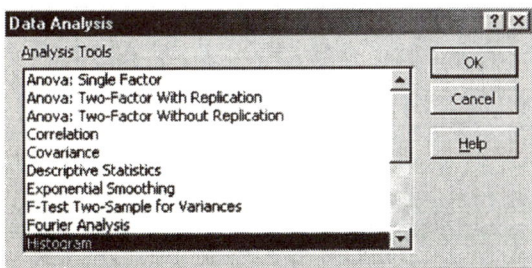

Data Analysis   ? X

Analysis Tools

Anova: Single Factor
Anova: Two-Factor With Replication
Anova: Two-Factor Without Replication
Correlation
Covariance
Descriptive Statistics
Exponential Smoothing
F-Test Two-Sample for Variances
Fourier Analysis
Histogram

OK

Cancel

Help

8. Complete the fields in the histogram dialog box as shown below. Click **OK**.

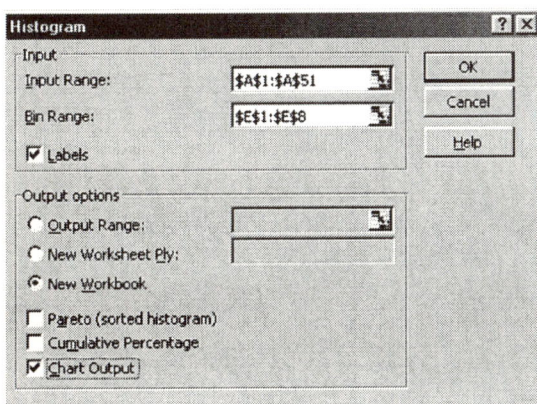

Histogram   ? X

Input

Input Range: $A$1:$A$51

Bin Range: $E$1:$E$8

☑ Labels

OK

Cancel

Help

Output options

○ Output Range:

○ New Worksheet Ply:

◉ New Workbook

☐ Pareto (sorted histogram)
☐ Cumulative Percentage
☑ Chart Output

9. You will now follow steps to modify the histogram so that it is presented in a more informative manner. Begin by making the chart taller so that it is easier to read. To do this, first click within the figure near a border. Black square handles appear. Click on the center handle at the bottom border of the figure and drag it down a few rows.

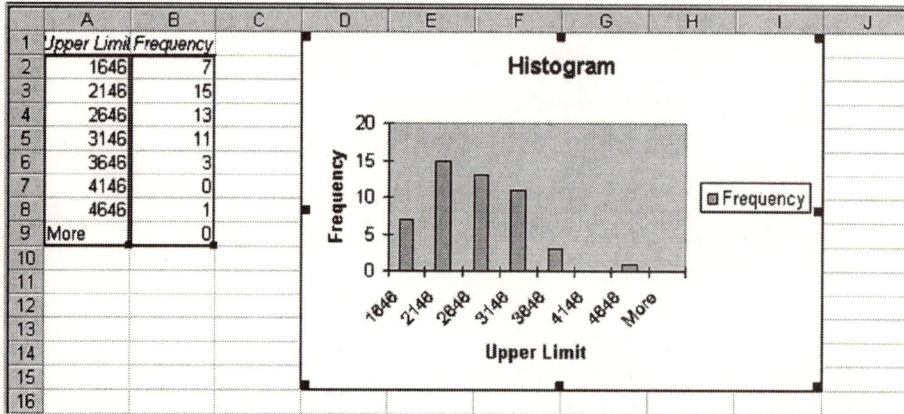

12. Remove the "More" category from the X axis. **Right click** in the gray plot area of the histogram and select **Source Data** from the shortcut menu that appears.

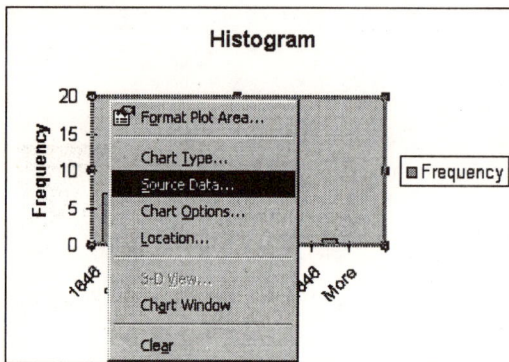

13. Click the **Series** tab at the top of the Source Data dialog box. "More" appears in cell A9 of the worksheet and its zero frequency appears in cell B9. To exclude the information in row 9 from the chart, edit the entry in the Values window and the entry in the Category (X) axis labels window. The entry in the Values window should read **=Sheet1!$B$2:$B$8**. The entry in the Category (X) axis labels window should read **=Sheet1!$A$2:$A$8**. Click **OK**.

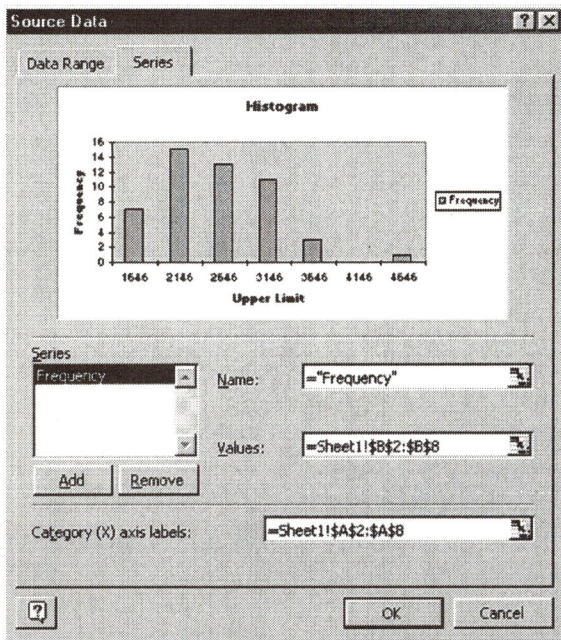

14. **Right click** in the gray plot area of the histogram and select **Chart Options** from the menu that appears.

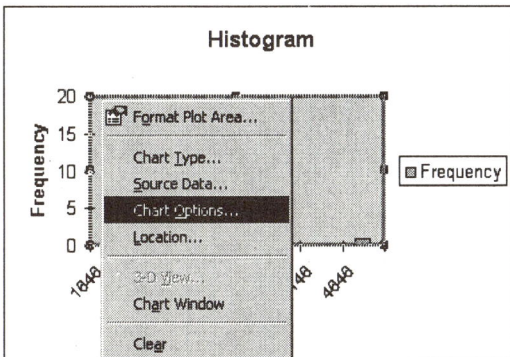

15.   Click the **Titles** tab at the top of the Chart Options dialog box. In the Chart title field, replace "Histogram" with **Monthly Milk Production of 50 Holstein Dairy Cows**. In the Category (X) axis field, enter **Pounds**.

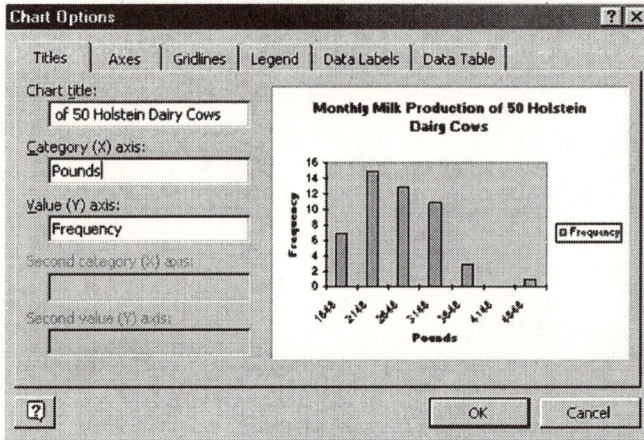

16.   Click the **Gridlines** tab at the top of the Chart Options dialog box. Click in the box next to **Major gridlines** under Value (Y) axis so that a checkmark appears there.

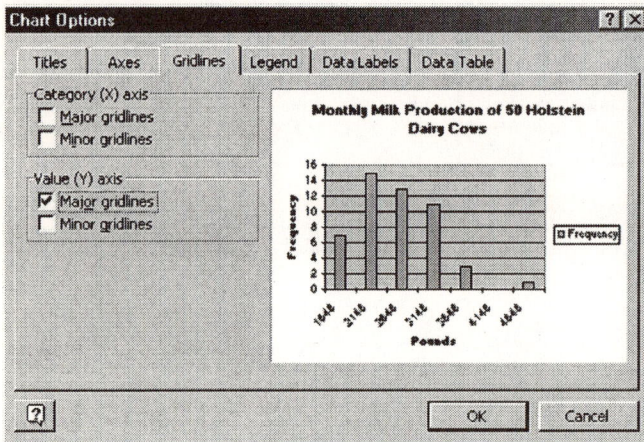

17. Click the **Legend** tab at the top of the Chart Options dialog box. Click in the box to the left of **Show Legend** to remove the checkmark. The removal of the checkmark will delete the frequency legend from the right side of the histogram chart. Click **OK**.

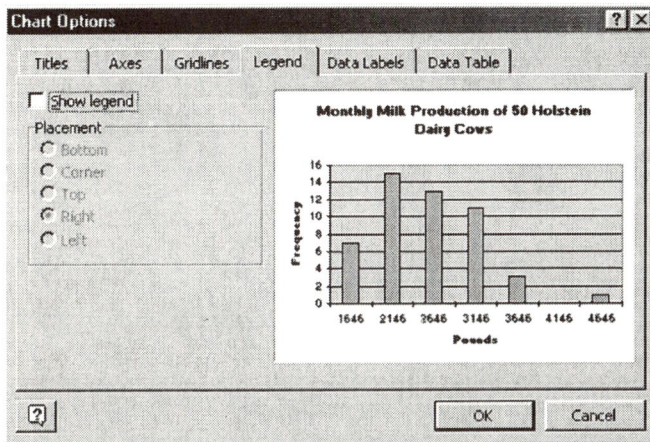

18. Remove the space between the vertical bars. **Right click** on one of the vertical bars. Select **Format Data Series** from the shortcut menu that appears.

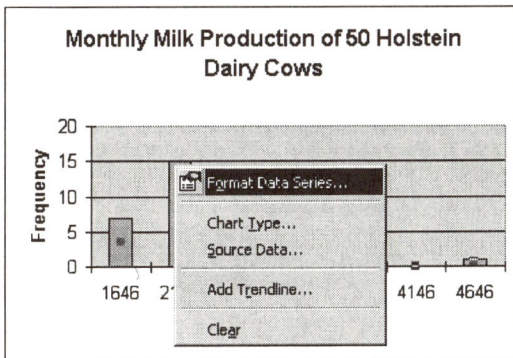

19. Click on the **Options** tab at the top of the Format Data Series dialog box. Change the value in the Gap width box to 0. Click **OK**.

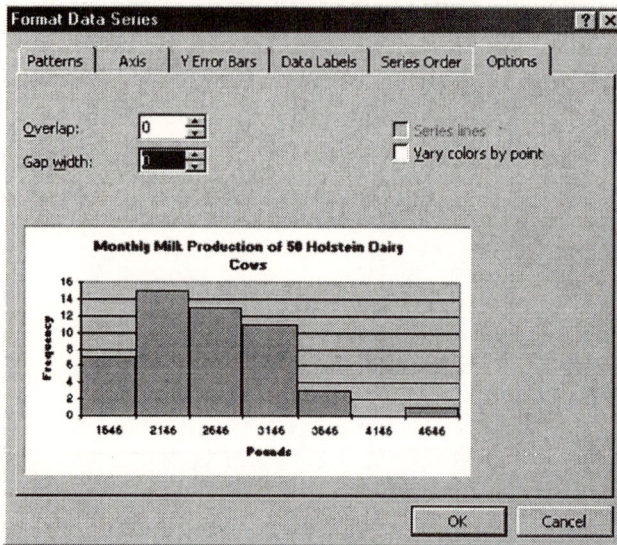

Your histogram should look similar to the one displayed below.

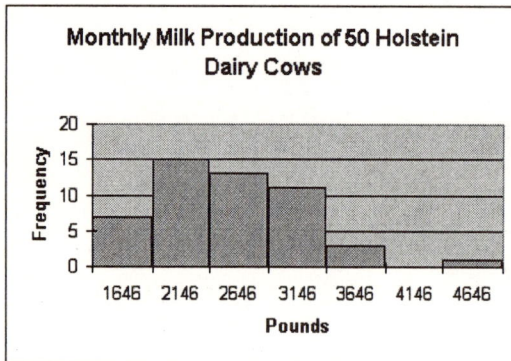

Probability

Section 3.2

▶ Exercise 26 (pg. 129)	Simulating the "Birthday Problem"

You will be generating 24 random numbers between 1 and 365.

> *If the PHStat2add-in has not been loaded, you will need to load it before continuing.*
> *Follow the instructions in Section GS 8.2.*

1. Open a new Excel worksheet.

2. You will be entering the numbers 1 through 365 in the worksheet. To do this, you will fill column A with a number series. Begin by entering the numbers 1 and 2 in column A as shown below. You are starting a number series that will increase in increments of one.

	A	B	C	D	E	F	G
1	1						
2	2						

3. You will now complete the series all the way to 365. Begin by clicking in cell A1 and dragging down to cell A2 so that both cells are highlighted as shown below.

	A	B	C	D	E	F	G
1	1						
2							

4. Move the mouse pointer in cell A2 to the bottom right corner of that cell so that the white plus sign turns into a black plus sign. This black plus sign is called the "fill handle." While the fill handle is displayed, hold down on the left mouse button and

drag down column A until you reach cell A365. Release the mouse button. You should see the numbers from 1 to 365 in column A.

	A	B	C	D	E	F	G
359	359						
360	360						
361	361						
362	362						
363	363						
364	364						
365	365						

5.   At the top of the screen, click **Tools** → **Data Analysis**. Select **Sampling** and click **OK**.

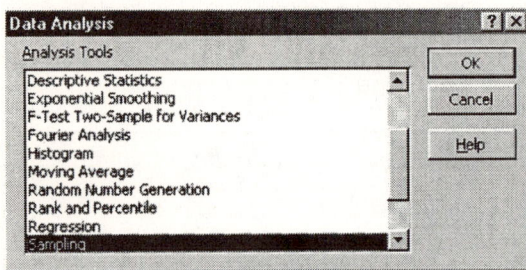

Data Analysis

Analysis Tools

- Descriptive Statistics
- Exponential Smoothing
- F-Test Two-Sample for Variances
- Fourier Analysis
- Histogram
- Moving Average
- Random Number Generation
- Rank and Percentile
- Regression
- Sampling

[OK] [Cancel] [Help]

6.   Complete the Sampling dialog box as shown below. Click **OK**.

You will be generating 10 different samples of 24 numbers. The first set of 24 numbers will be placed in column B, the second in column C, etc.

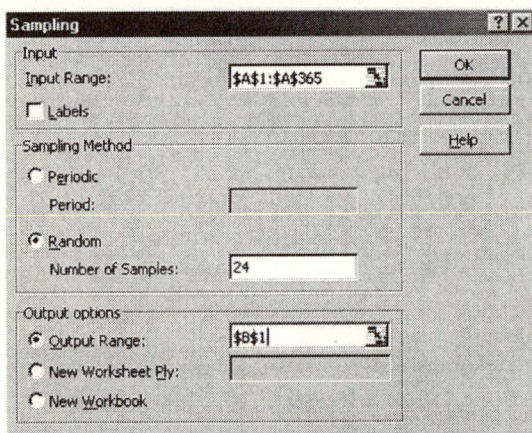

Sampling

Input
  Input Range: $A$1:$A$365
  ☐ Labels

Sampling Method
  ○ Periodic
    Period:
  ● Random
    Number of Samples: 24

Output options
  ● Output Range: $B$1
  ○ New Worksheet Ply:
  ○ New Workbook

[OK] [Cancel] [Help]

7. The 24 numbers, generated randomly with replacement, are displayed in column B. (Because the numbers were generated randomly, it is not likely that your output will be exactly the same.) Construct a frequency distribution to see if there are any repetitions. Select **PHStat → Descriptive Statistics → One-Way Tables & Charts**.

	A	B	C	D	E	F	G
1	1	308					
2	2	273					
3	3	120					
4	4	57					
5	5	350					
6	6	276					
7	7	259					
8	8	28					
9	9	51					
10	10	308					
11	11	7					
12	12	237					
13	13	60					
14	14	107					
15	15	171					
16	16	265					
17	17	353					
18	18	172					
19	19	211					
20	20	201					
21	21	89					
22	22	259					

8. Complete the One-Way Tables & Charts dialog box as shown below. Click **OK**.

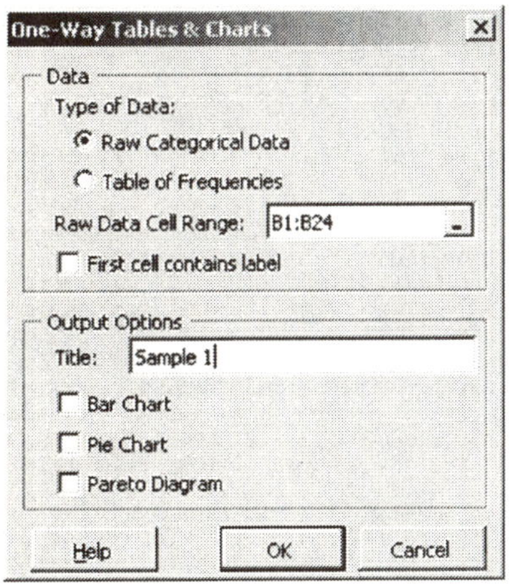

9.   The frequency distribution table is displayed in a worksheet named "OneWayTable." In this example, you can see that the value of 259 has a frequency of 2.

	A	B
1	Sample 1	
2		
3	Count of Variable	
4	Variable	Total
5	7	1
6	28	1
7	51	1
8	57	1
9	60	1
10	89	1
11	107	1
12	120	1
13	171	1
14	172	1
15	194	1
16	201	1
17	211	1
18	237	1
19	259	2
20	265	1
21	273	1
22	276	1

Scrolling down, you can see that, in this example, the value of 308 also has a frequency of 2.

23	308	2
24	345	1
25	350	1
26	353	1
27	Grand Total	24

10. Generate the second sample. Go back to the sheet containing the numbers 1 through 365 by clicking on the **Sheet1** tab near the bottom of the screen. Then click **Tools → Data Analysis**. Select **Sampling** and click **OK**. Complete the Sampling dialog box as shown below. The second set of randomly generated numbers will be placed in column C. Click **OK**.

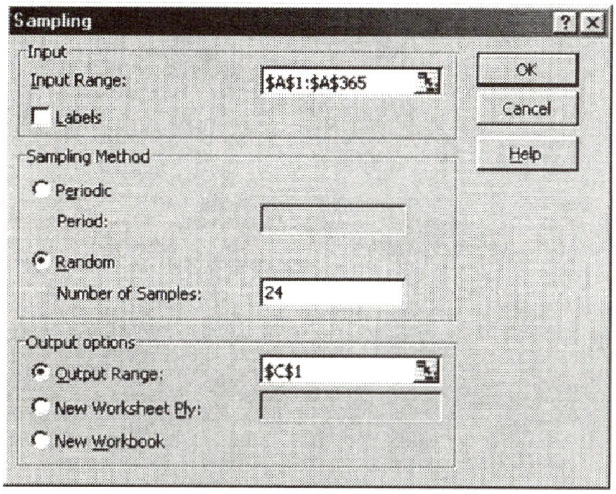

11. Construct a frequency distribution to see if there are any repetitions. Select **PHStat → One-Way Tables & Charts**. Complete the One-Way Tables & Charts dialog box as shown below. Click **OK**.

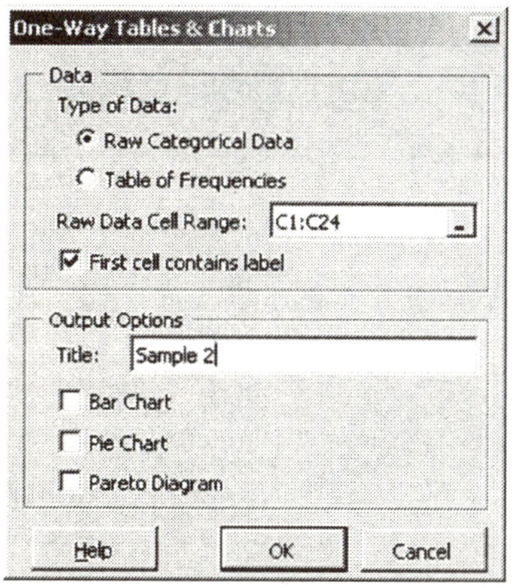

12. The frequency distribution table is displayed in a worksheet named "OneWayTable2." In this example, you can see that the value of 119 has a frequency of 2.

	A	B
1	Sample 2	
2		
3	Count of Variable	
4	Variable	Total
5	4	1
6	15	1
7	72	1
8	86	1
9	99	1
10	111	1
11	114	1
12	119	2
13	120	1
14	124	1
15	205	1
16	218	1
17	232	1
18	244	1
19	248	1
20	258	1
21	309	1
22	312	1

Scrolling down, you see no more repetitions in this example.

23	318	1
24	320	1
25	324	1
26	332	1
27	364	1
28	Grand Total	24

13. Repeat the appropriate steps until you have generated 10 different samples of 24 numbers.

Section 3.4

▶ Example 3 (pg. 142)	Finding the Number of Permutations of n Objects

1. Open a new Excel worksheet and click in cell **A1** to place the output there. You will be using the PERMUT function to calculate the number of permutations.

2. Click **Insert → Function**.

3. Under Function category, select **Statistical** by clicking on it. Under Function name, select **PERMUT**. Click **OK**.

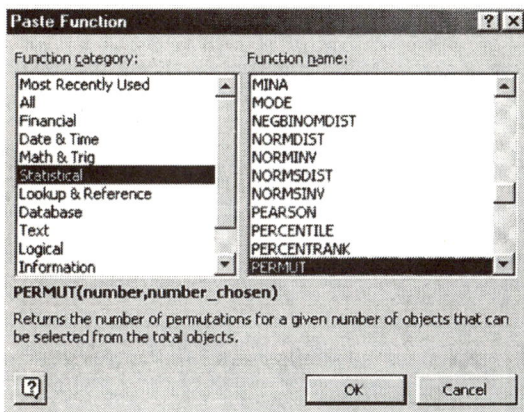

4. Complete the PERMUT dialog box as shown below. Click **OK**.

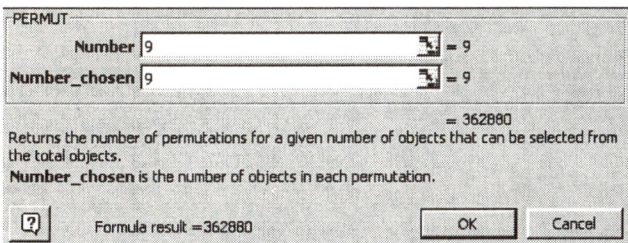

The output is shown below. There are 362,880 different batting orders.

	A	B	C	D	E	F	G
1	362880						
2							

◄

► Example 7 (pg. 145)	Finding the Number of Combinations

1. Open a new Excel worksheet and click in cell **A1** to place the output there. You will be using the COMBIN function to calculate the number of combinations.

2. Click **Insert** → **Function**.

3. Under Function category, select **Math & Trig** by clicking on it. Under Function name, select **COMBIN**. Click **OK**.

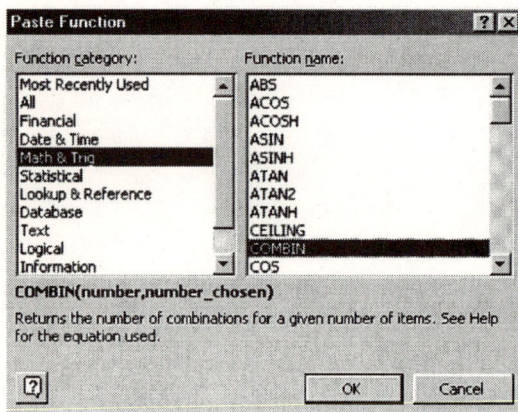

4. Complete the COMBIN dialog box as shown below. Click **OK**.

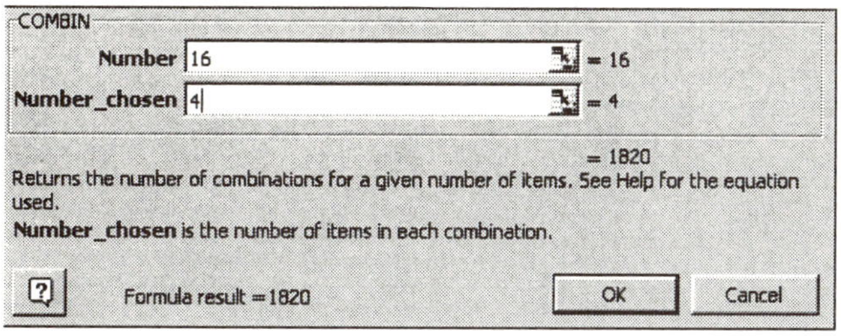

```
┌─COMBIN────────────────────────────────────────────────────┐
│         Number │16                        │ 🔢 = 16         │
│  Number_chosen │4│                        │ 🔢 = 4          │
│                                             = 1820          │
│  Returns the number of combinations for a given number of items. See Help for the equation │
│  used.                                                      │
│  Number_chosen is the number of items in each combination.  │
│                                                             │
│  ❓    Formula result = 1820          │  OK  │  Cancel  │   │
└─────────────────────────────────────────────────────────────┘
```

The output is shown below. There are 1,820 different combinations.

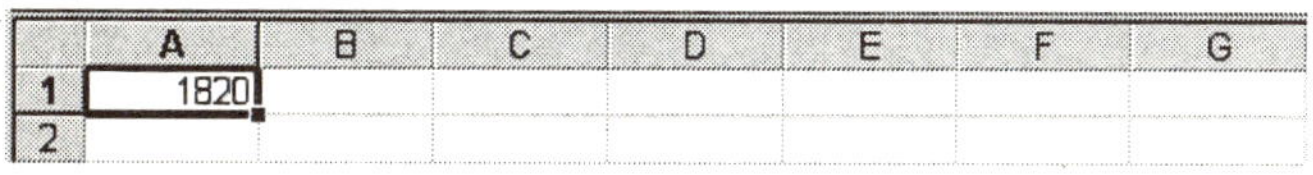

	A	B	C	D	E	F	G
1	1820						
2							

◀

Technology Lab

▶ Exercise 3 (pg. 151)	Selecting Randomly a Number from 1 to 11

If the PHStat2 add-in has not been loaded, you will need to load it before continuing. Follow the instructions in Section GS 8.2.

You will be given the steps to follow to complete Exercise 3 and Exercise 3B.

1. You will first select randomly a number between 1 and 11. Open a new Excel worksheet. Enter the numbers 1 through 11 in cells A1 through A11 as shown below.

	A	B	C	D	E	F	G
1	1						
2	2						
3	3						
4	4						
5	5						
6	6						
7	7						
8	8						
9	9						
10	10						
11	11						

2. At the top of the screen, click **Tools** → **Data Analysis**. Select **Sampling**, and click **OK**.

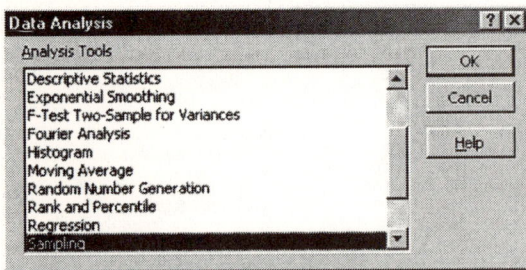

```
Data Analysis                          ? X
Analysis Tools
  Descriptive Statistics        ▲      [ OK ]
  Exponential Smoothing
  F-Test Two-Sample for Variances      [ Cancel ]
  Fourier Analysis
  Histogram                            [ Help ]
  Moving Average
  Random Number Generation
  Rank and Percentile
  Regression
  Sampling                      ▼
```

3. Complete the Sampling dialog box as shown below. Click **OK**.

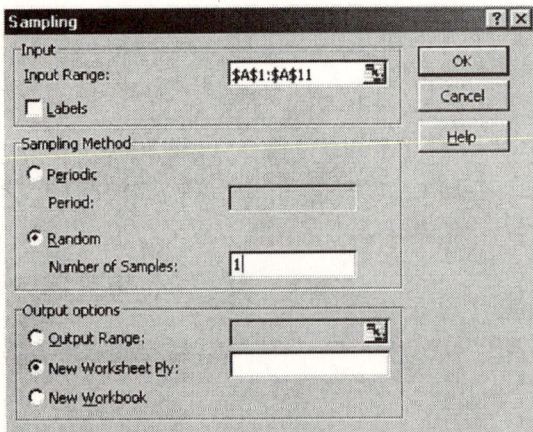

```
Sampling                               ? X
Input
  Input Range:      $A$1:$A$11         [ OK ]
  ☐ Labels                             [ Cancel ]

Sampling Method                        [ Help ]
  ○ Periodic
    Period:        [          ]

  ● Random
    Number of Samples:  1

Output options
  ○ Output Range:   [          ]
  ● New Worksheet Ply:  [          ]
  ○ New Workbook
```

4. The output is displayed in a new sheet. The number 2 was selected. Because the number was selected, your output may not be the same. You will now select randomly 100 integers from 1 to 11. Click the **Sheet1** tab near the bottom of the screen to go back to the worksheet displaying the numbers from 1 to 11.

	A	B	C	D	E	F	G
1	2						
2							

5. Click **Tools → Data Analysis**. Select **Sampling** and click **OK**.

6. Complete the Sampling dialog box as shown below. Click **OK**.

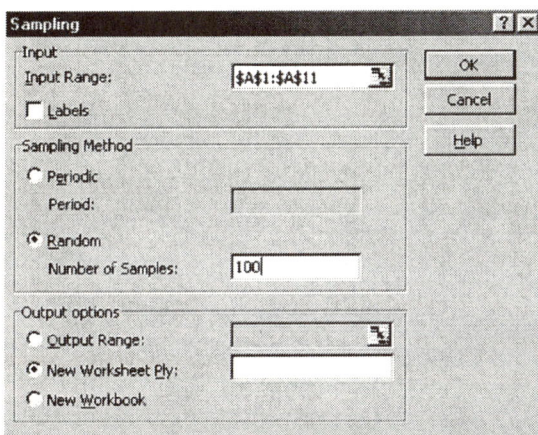

7. The output is displayed in a new sheet. You will need to scroll down to see the entire listing of the 100 randomly generated integers. You will now tally the results. Select **PHStat → Descriptive Statistics → One-Way Tables & Charts**.

	A	B	C	D	E	F	G
1	4						
2	1						
3	7						
4	2						
5	8						
6	8						
7	2						

8.   Complete the One-Way Tables & Charts dialog box as shown below. Click **OK**.

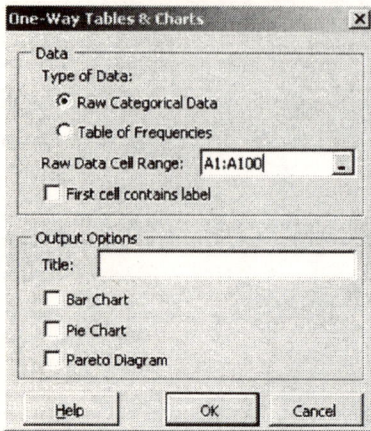

```
One-Way Tables & Charts                    [X]
  ┌─ Data ──────────────────────────────────┐
  │   Type of Data:                         │
  │      (•) Raw Categorical Data           │
  │      ( ) Table of Frequencies           │
  │   Raw Data Cell Range:  [A1:A100]  [.]  │
  │      [ ] First cell contains label      │
  └─────────────────────────────────────────┘
  ┌─ Output Options ────────────────────────┐
  │   Title: [                            ] │
  │                                         │
  │      [ ] Bar Chart                      │
  │      [ ] Pie Chart                      │
  │      [ ] Pareto Diagram                 │
  └─────────────────────────────────────────┘
     [ Help ]      [  OK  ]      [ Cancel ]
```

The output for Exercise 3B is displayed below. Your output should have the same format. Because the numbers were generated randomly, however, it is not likely that your numbers will be exactly the same.

	A	B	C
1	One-Way Summary Table		
2			
3	Count of Variable		
4	Variable ▾	Total	
5	1	7	
6	2	5	
7	3	6	
8	4	14	
9	5	13	
10	6	4	
11	7	8	
12	8	8	
13	9	20	
14	10	7	
15	11	8	
16	Grand Total	100	

◄

▶ **Exercise 5 (pg. 151)**   Composing Mozart Variations with Dice

---

If the PHStat2 add-in has not been loaded, you will need to load it before continuing. Follow the instructions in Section GS 8.2.

You will first be given the steps to follow to complete Exercise 5 and then the steps to follow to complete Exercise 5B. In Exercise 5, you will select randomly two integers from 1, 2, 3, 4, 5, and 6. This simulates tossing two six-sided dice one time. You will then sum the two integers and subtract 1 from the sum. This is Mozart's procedure (described at the top of page 151) for selecting a musical phrase from 11 different choices.

1. To select randomly two integers between 1 and 6, begin by opening a new Excel worksheet and entering the numbers 1 through 6 in column A as shown below.

	A	B	C	D	E	F	G
1	1						
2	2						
3	3						
4	4						
5	5						
6	6						

2. At the top of the screen, click **Tools → Data Analysis**. Select **Sampling** and click **OK**.

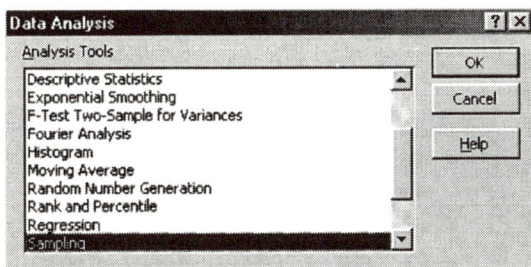

Data Analysis

Analysis Tools

Descriptive Statistics
Exponential Smoothing
F-Test Two-Sample for Variances
Fourier Analysis
Histogram
Moving Average
Random Number Generation
Rank and Percentile
Regression
Sampling

OK
Cancel
Help

3. Complete the Sampling dialog box as shown below. Click **OK**.

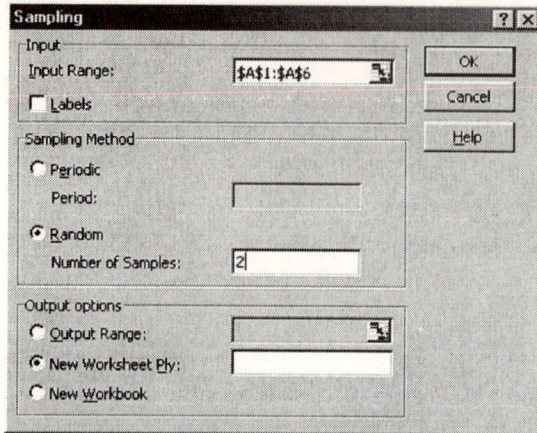

4. The output is displayed in a new worksheet. The numbers 5 and 3 were selected.
    Mozart's instructions are to sum the numbers and subtract 1. The result is 7.
    Therefore, musical phrase number 7 would be selected for the first bar of Mozart's
    minuet.

	A	B	C	D	E	F	G
1	5						
2	3						

5. Exercise 5B asks you to select 100 integers between 1 and 11. This simulates, 100
    times, Mozart's procedure of tossing two die, finding the sum, and subtracting 1.
    You are also asked to tally the results. Begin by entering the numbers 1 through 11
    in a new worksheet as shown below.

	A	B	C	D	E	F	G
1	1						
2	2						
3	3						
4	4						
5	5						
6	6						
7	7						
8	8						
9	9						
10	10						
11	11						

6. Click **Tools → Data Analysis**. Select **Sampling** and click **OK**.

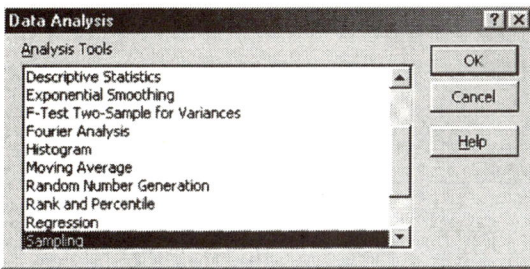

7. Complete the Sampling dialog box as shown below. Click **OK**.

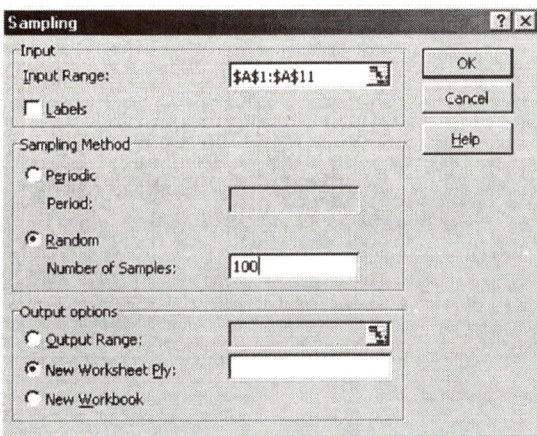

8. The output is displayed in a new sheet. You need to scroll down to see the entire listing of the 100 randomly generated integers. You will now tally the results. Select **PHStat → Descriptive Statistics → One-Way Tables & Charts**.

9. Complete the One-Way Tables and Charts dialog box as shown below. Click **OK**.

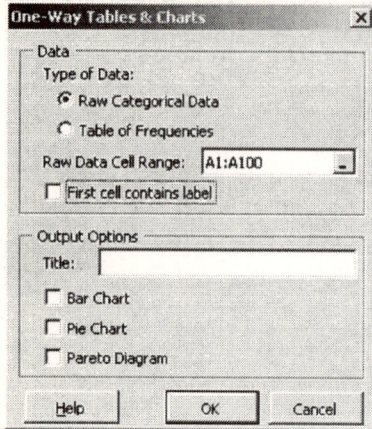

```
┌─────────────────────────────────────────┐
│ One-Way Tables & Charts            ×│
│ ┌─ Data ──────────────────────────────┐ │
│ │  Type of Data:                       │ │
│ │    ⦿ Raw Categorical Data            │ │
│ │    ○ Table of Frequencies            │ │
│ │  Raw Data Cell Range:  A1:A100    ▪  │ │
│ │    □ First cell contains label       │ │
│ └──────────────────────────────────────┘ │
│ ┌─ Output Options ────────────────────┐ │
│ │  Title: [                         ]  │ │
│ │    □ Bar Chart                       │ │
│ │    □ Pie Chart                       │ │
│ │    □ Pareto Diagram                  │ │
│ └──────────────────────────────────────┘ │
│   [ Help ]     [ OK ]     [ Cancel ]     │
└─────────────────────────────────────────┘
```

10. The output from Exercise 5B is displayed below. Because the numbers were generated randomly, it unlikely that your numbers will be exactly the same.

	A	B	C
1	One-Way Summary Table		
2			
3	Count of Variable		
4	Variable ▾	Total	
5	1	9	
6	2	8	
7	3	13	
8	4	5	
9	5	9	
10	6	10	
11	7	4	
12	8	9	
13	9	13	
14	10	10	
15	11	10	
16	Grand Total	100	

Discrete Probability Distributions

4

Section 4.2

▶ Example 5 (pg. 179) Finding a Binomial Probability

1. Open a new Excel worksheet and click in cell **A1** to place the output there.

2. Click **Insert → Function**.

3. Under Function category, select **Statistical**. Under Function name, select **BINOMDIST**. Click **OK**.

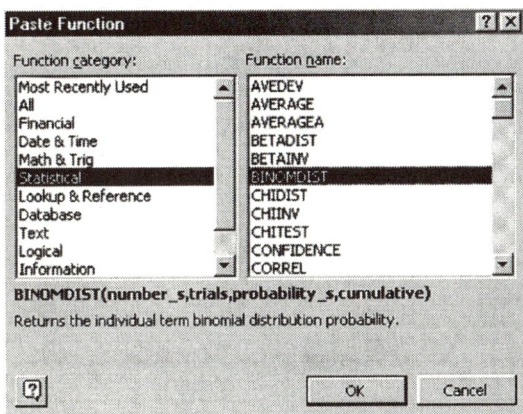

Paste Function

Function category:
Most Recently Used
All
Financial
Date & Time
Math & Trig
Statistical
Lookup & Reference
Database
Text
Logical
Information

Function name:
AVEDEV
AVERAGE
AVERAGEA
BETADIST
BETAINV
BINOMDIST
CHIDIST
CHIINV
CHITEST
CONFIDENCE
CORREL

BINOMDIST(number_s,trials,probability_s,cumulative)
Returns the individual term binomial distribution probability.

OK   Cancel

4.   Complete the BINOMDIST dialog box as shown below. Click **OK**.

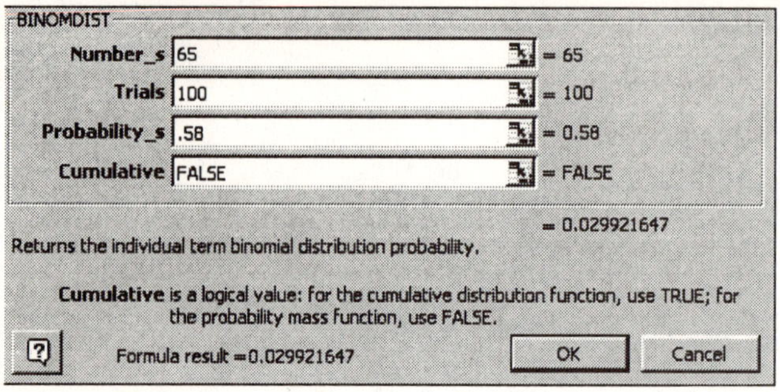

```
┌─ BINOMDIST ────────────────────────────────────────────────┐
│        Number_s │65                              │ ▚ = 65   │
│          Trials │100                             │ ▚ = 100  │
│     Probability_s │.58                           │ ▚ = 0.58 │
│        Cumulative │FALSE                         │ ▚ = FALSE│
│                                                              │
│                                        = 0.029921647         │
│  Returns the individual term binomial distribution probability.│
│                                                              │
│     Cumulative is a logical value: for the cumulative distribution function, use TRUE; for │
│              the probability mass function, use FALSE.        │
│                                                              │
│   [?]      Formula result =0.029921647        [   OK   ] [ Cancel ]│
└──────────────────────────────────────────────────────────────┘
```

The BINOMDIST function returns a result of 0.029922.

	A	B	C	D	E	F	G
1	0.029922						
2							

◀

► Example 7 (pg. 181)	Constructing and Graphing a Binomial Distribution

1.   Open a new, blank Excel worksheet and enter the information shown below. You will be using the BINOMDIST function to calculate binomial probabilities for 0, 1, 2, 3, 4, 5, and 6 households.

	A	B	C	D	E	F	G
1	Households	Relative frequency					
2	0						
3	1						
4	2						
5	3						
6	4						
7	5						
8	6						

2.   Click in cell **B2** below "Relative frequency." Click **Insert → Function**.

3. Under Function category, select **Statistical**. Under Function name, select
    BINOMDIST. Click **OK**.

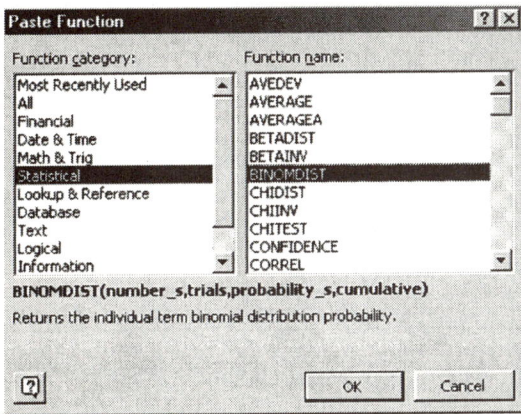

Paste Function	? X
Function category:	Function name:
Most Recently Used All Financial Date & Time Math & Trig Statistical Lookup & Reference Database Text Logical Information	AVEDEV AVERAGE AVERAGEA BETADIST BETAINV BINOMDIST CHIDIST CHIINV CHITEST CONFIDENCE CORREL

BINOMDIST(number_s,trials,probability_s,cumulative)

Returns the individual term binomial distribution probability.

[?]   OK   Cancel

4. Complete the BINOMDIST dialog box as shown below. Click **OK**.

---

*You are entering a relative cell address without dollar signs (i.e., A2) in the
Number_s field because you will be copying the contents of cell B2 to cells B3
through B8. You want the column A cell address to change from A2 to A3, A4, A5,
..., A8 when the formula is copied from cell B2 to cells B3 through B8.*

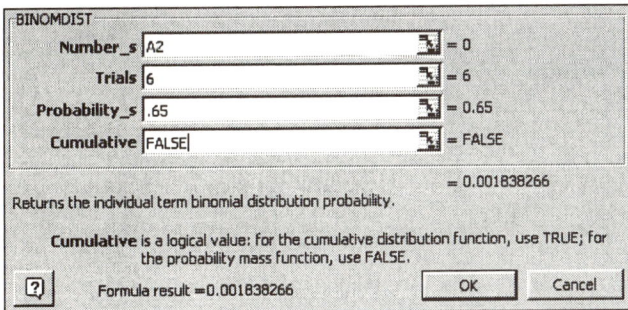

BINOMDIST		
Number_s	A2	= 0
Trials	6	= 6
Probability_s	.65	= 0.65
Cumulative	FALSE	= FALSE

= 0.001838266

Returns the individual term binomial distribution probability.

Cumulative is a logical value: for the cumulative distribution function, use TRUE; for
the probability mass function, use FALSE.

[?]   Formula result =0.001838266   OK   Cancel

5. Copy the contents of cell B2 to cells B3 through B8.

	A	B	C	D	E	F	G
1	Households	Relative frequency					
2	0	0.001838					
3	1	0.020484					
4	2	0.095102					
5	3	0.235491					
6	4	0.328005					
7	5	0.243661					
8	6	0.075419					

6. Click in any cell in the data table. Then click **Insert** → **Chart**.

7. Under Chart type, in the Chart Type dialog box, select **Column**. Under Chart sub-type, select the first one in the top row by clicking on it. Click **Next**.

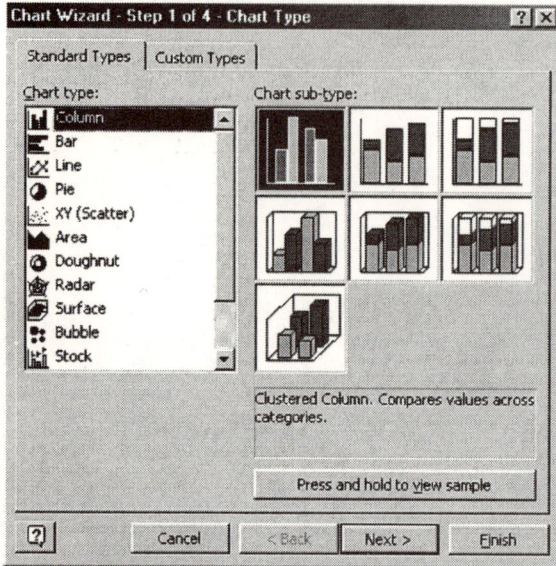

8. Check the accuracy of the data range in the Chart Source Data dialog box. It should read **=Sheet1!$A$1:$B$8**. Make any necessary corrections. Then click **Next**.

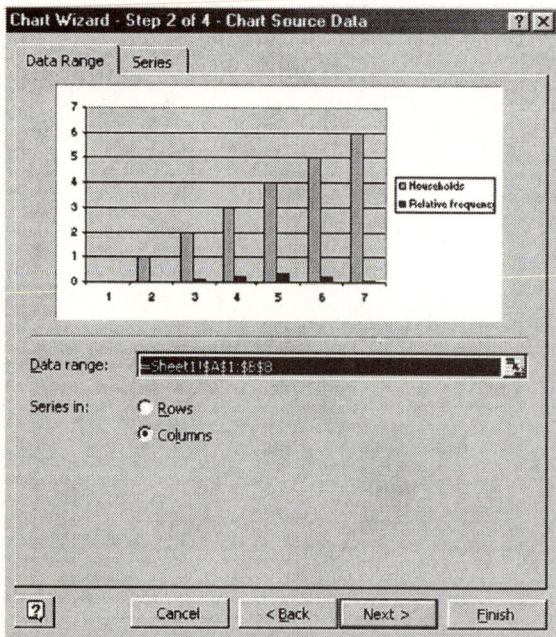

9. Click the **Titles** tab at the top of the Chart Options dialog box. In the Chart title window, enter **Subscribing to Cable TV**. In the Category (X) axis window, enter **Households**. In the Value (Y) axis window, enter **Relative Frequency**.

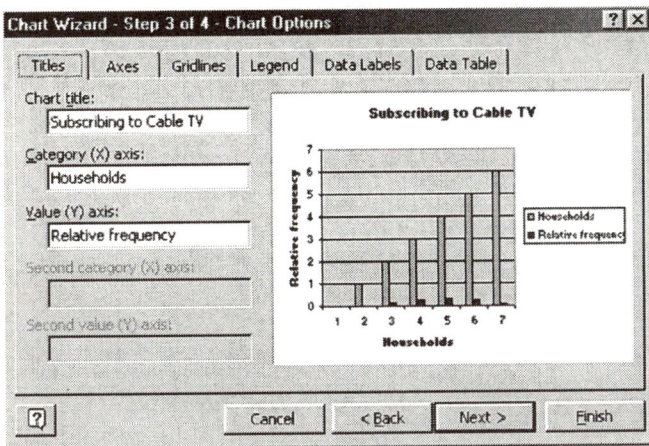

10. Click the **Legend** tab at the top of the Chart Options dialog box. Click in the box to the left of **Show legend** so that no checkmark appears there. This will remove the Households and Relative frequency legends on the right side of the chart. Click **Next**.

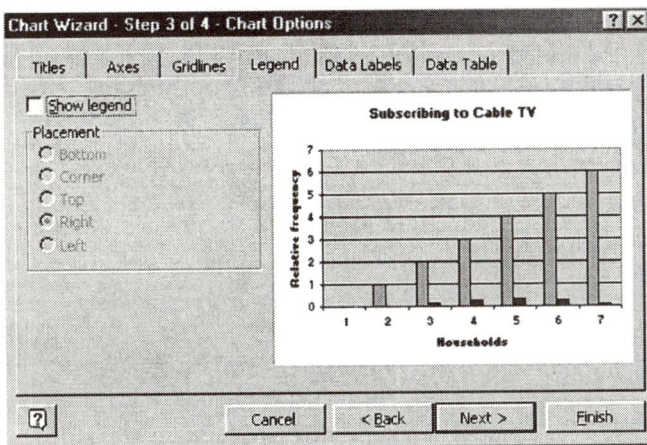

11. The Chart Location dialog box presents two options for placement of the chart. For this example, select **As object in**. Click **Finish**.

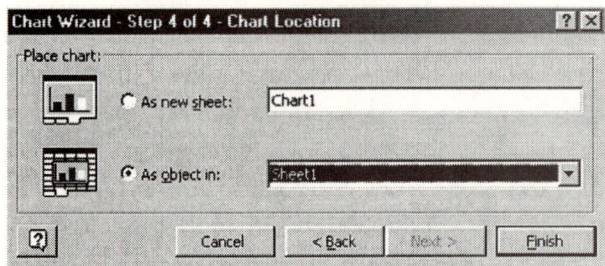

12. Make the chart taller so that it is easier to read. To do this, first click within the figure near a border so that black square handles appear. Click on the center handle on the bottom border of the figure and drag it down a few rows. Your chart should look similar to the one shown below.

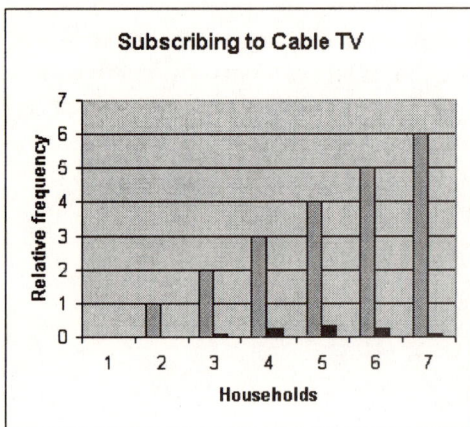

13. Correct the number scales displayed on the Y-axis and the X-axis. **Right click** on one of the vertical bars. Select **Source Data** from the shortcut menu that appears.

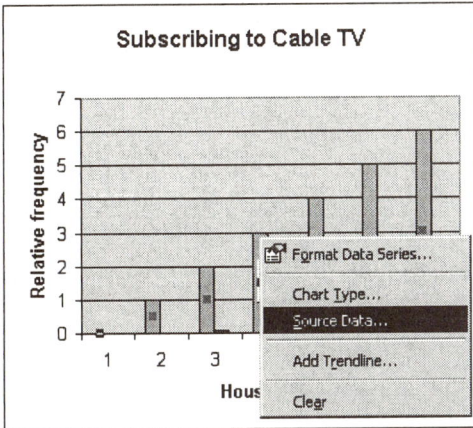

14. Click the **Series** tab at the top of the Source Data dialog box. In the Series window, select **Relative Frequency** by clicking on it. Then click the **Remove** button below the Series window.

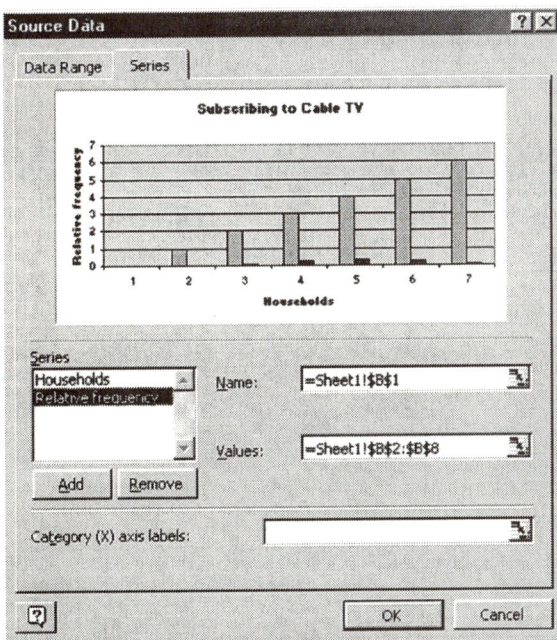

15. The Y-axis values are displayed in column B, so the entry in the **Values** window should be **=Sheet1!$B$2:$B$8**. First delete the information in the Values window. Next, click in the Values window, click on cell B2 of the worksheet, and drag down to cell B8. In the Values window, you should now see **=Sheet1!$B$2:$B$8**.

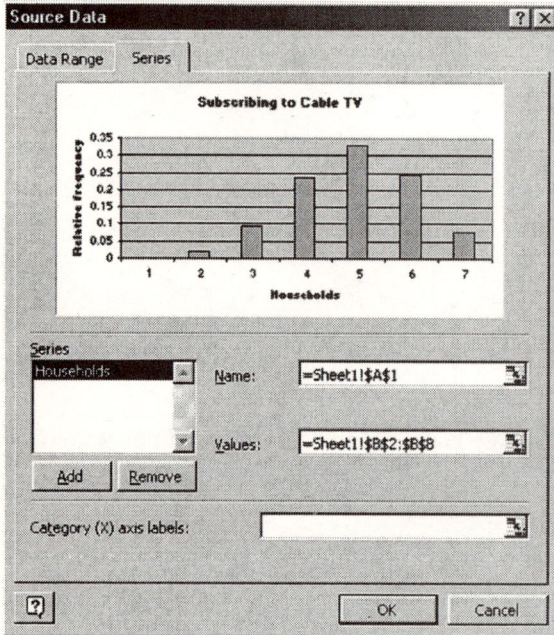

16. Column A contains the numbers you want displayed on the X-axis. Click in the Category (X) axis labels window. Then click on cell A2 of the worksheet and drag down to cell A8. The entry in the Category (X) axis labels field should now read **=Sheet1!$A$2:$A$8**. Click **OK**.

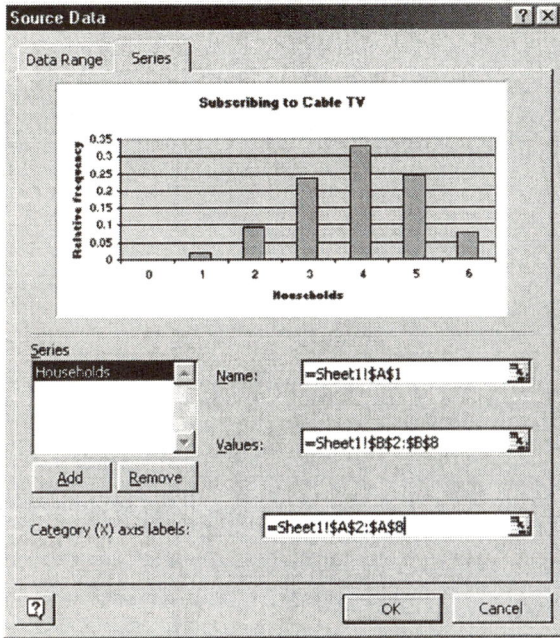

17. Remove the space between the vertical bars. **Right click** on one of the vertical bars. Select **Format Data Series** from the shortcut menu that appears.

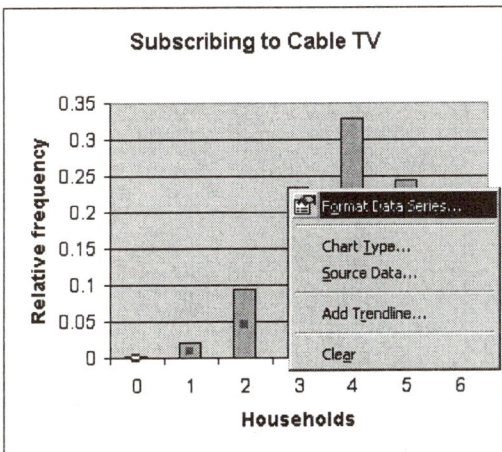

18. Click the **Options** tab at the top of the Format Data Series dialog box. Change the value in the Gap width box to 0. Click **OK**. Your relative frequency histogram should look similar to the one shown below.

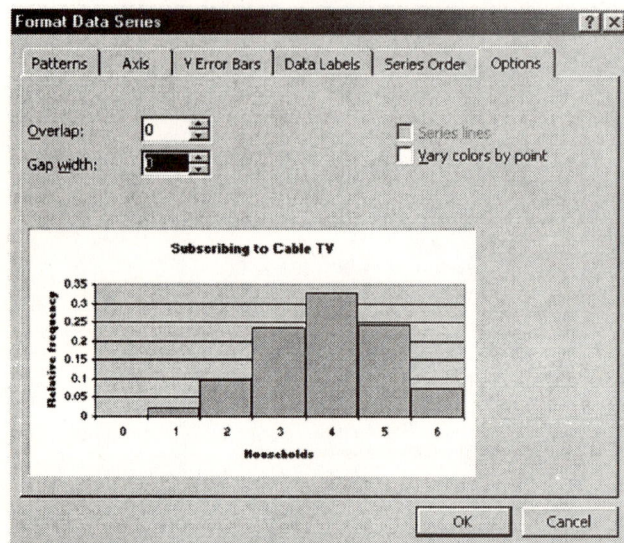

Section 4.3

| ▶ Example 3 (pg. 190) | Finding a Poisson Probability |

You will be using the POISSON function to find the probability that 2 rabbits are found on any given acre of a field when a population count shows that there is an average of 3.6 rabbits per acre living in the field.

1. Open a new Excel worksheet and click in cell **A1** to place the output there.

2. At the top of the screen, click **Insert → Function**.

3. Under Function category, select **Statistical**. Under Function name, select
    POISSON. Click **OK**.

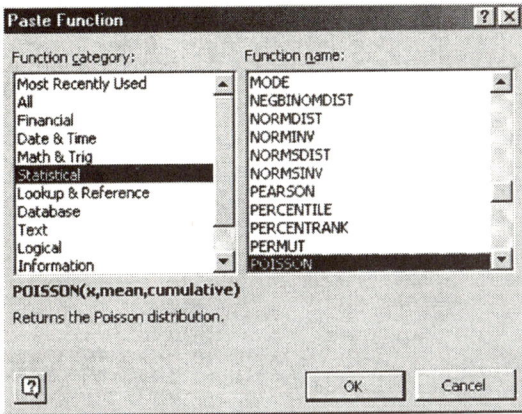

4. Complete the POISSON dialog box as shown below. Click **OK**.

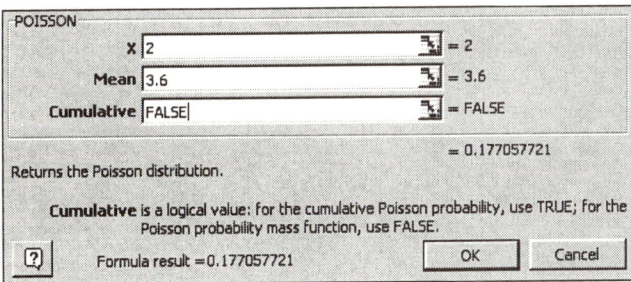

The POISSON function returns as result of 0.177058.

Technology Lab

▶ Exercise 1 (pg. 195)	Creating a Poisson Distribution with μ = 4 for x = 0 to 10

1. Open a new Excel worksheet and enter the numbers 0 through 10 in column A as shown below. Then click in cell **B1** of the worksheet to place the output from the POISSON function there.

	A	B	C	D	E	F	G
1	0						
2	1						
3	2						
4	3						
5	4						
6	5						
7	6						
8	7						
9	8						
10	9						
11	10						

2. At the top of the screen, click **Insert → Function**.

3. Under Function category, select **Statistical**. Under Function name, select **POISSON**. Click **OK**.

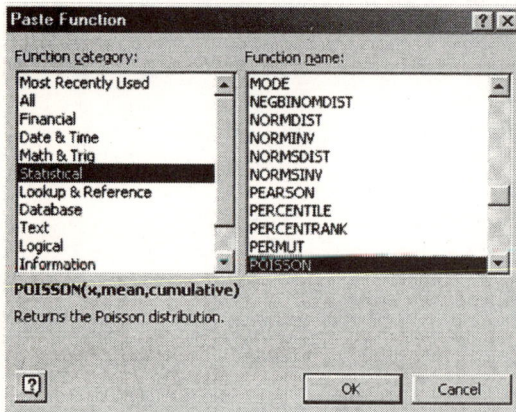

Paste Function	? X
Function category:	Function name:
Most Recently Used	MODE
All	NEGBINOMDIST
Financial	NORMDIST
Date & Time	NORMINV
Math & Trig	NORMSDIST
Statistical	NORMSINV
Lookup & Reference	PEARSON
Database	PERCENTILE
Text	PERCENTRANK
Logical	PERMUT
Information	POISSON

POISSON(x,mean,cumulative)

Returns the Poisson distribution.

[?]   OK   Cancel

4. Complete the POISSON dialog box as shown below. Click **OK**.

You are entering a relative cell address without dollar signs (i.e., A1) in the X field because you will be copying the contents of cell B1 to cells B2 through B11. You want the column A cell address to change from A1 to A2, A3, A4, ..., A11 when the formula is copied from cell B1 to cells B2 through B11.

```
┌─POISSON────────────────────────────────────────────────┐
│              X │A1                        │ ⁵⌐│ = 0     │
│         Mean │4                           │ ⁵⌐│ = 4     │
│    Cumulative │FALSE                      │ ⁵⌐│ = FALSE │
│                                          = 0.018315639  │
│  Returns the Poisson distribution.                      │
│                                                         │
│      Cumulative is a logical value: for the cumulative Poisson probability, use TRUE; for the │
│              Poisson probability mass function, use FALSE.                        │
│  ┌──┐                                     ┌──────┐  ┌────────┐ │
│  │ ? │   Formula result =0.018315639      │  OK  │  │ Cancel │ │
│  └──┘                                     └──────┘  └────────┘ │
└─────────────────────────────────────────────────────────┘
```

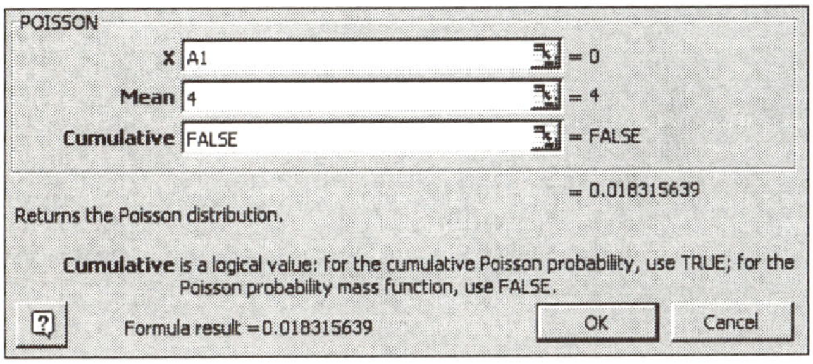

5. Copy the contents of cell B1 to cells B2 through B11.

	A	B	C	D	E	F	G
1	0	0.018316					
2	1	0.073263					
3	2	0.146525					
4	3	0.195367					
5	4	0.195367					
6	5	0.156293					
7	6	0.104196					
8	7	0.05954					
9	8	0.02977					
10	9	0.013231					
11	10	0.005292					

6. Construct a histogram of the Poisson distribution. First click on any cell of the table in the worksheet. Then, at the top of the screen, click **Insert → Chart**.

7. Under Chart type, in the Chart Type dialog box, select **Column**. Under Chart sub-type, select the first one in the top row by clicking on it. Click **Next**.

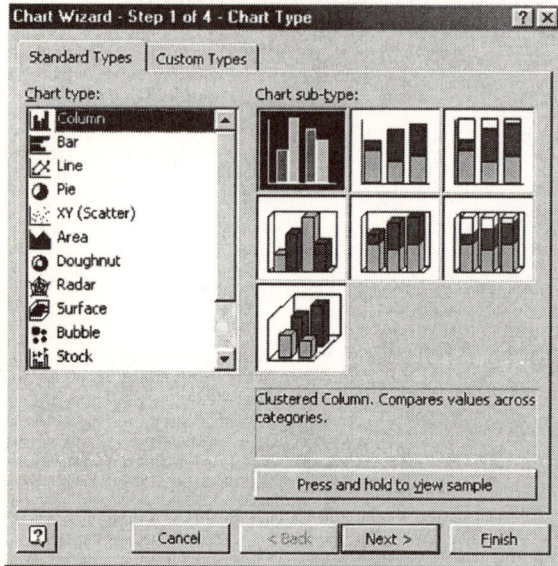

8. Check the accuracy of the data range in the Chart Source Data dialog box. It should read **=Sheet1!$A$1:$B$11**. Make any necessary corrections. Then click **Next**.

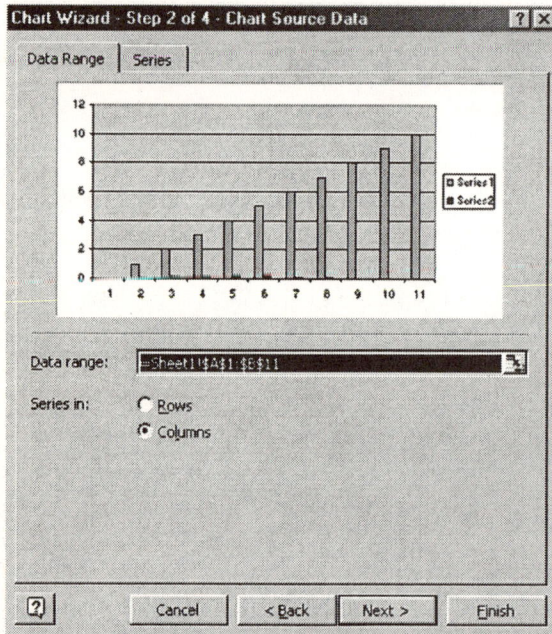

9. Click the **Titles** tab at the top of the Chart Options dialog box. In the Chart title window, enter **Customers Arriving at the Check-out Counter**. In the Category (X) axis window, enter **Number of Arrivals per Minute**. In the Value (Y) axis window, enter **Probability**.

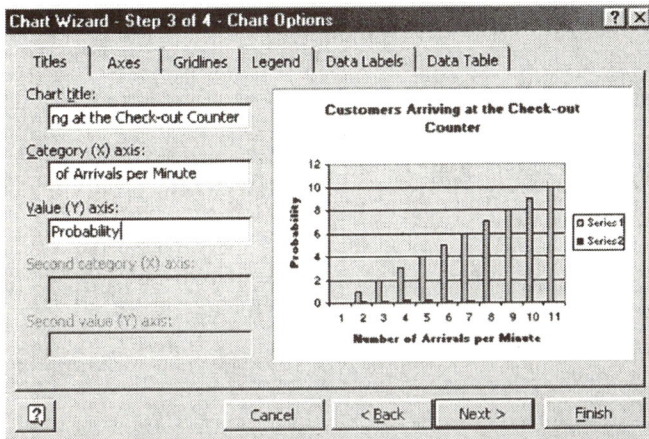

10. Click the **Legend** tab at the top of the Chart Options dialog box. Click in the box to the left of **Show Legend** to remove the checkmark. The removal of the checkmark will delete the Series 1 and Series 2 legends from the right side of the histogram chart. Click **Next**.

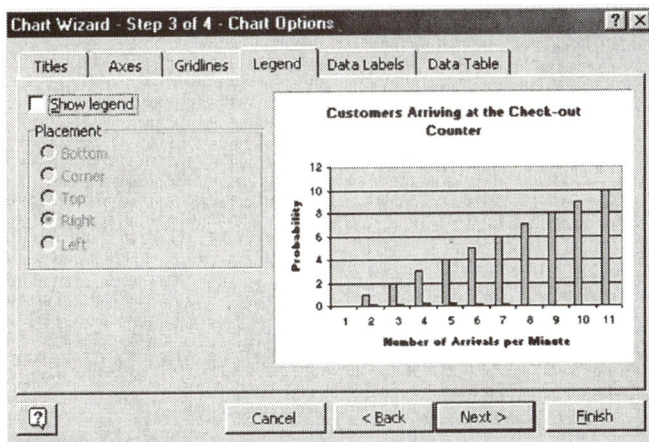

11. The Chart Location dialog box presents two options for placement of the chart. For this exercise, select **As object in**. Click **Finish**.

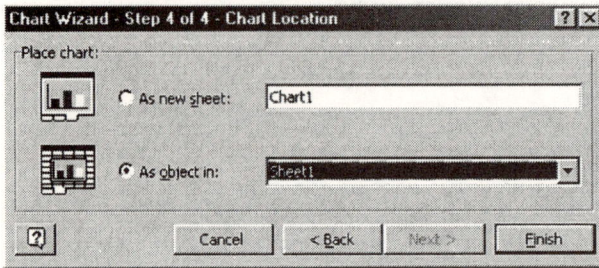

12. Make the chart taller so that it is easier to read. To do this, first click within the figure near a border so that black square handles appear. Click on the center handle on the bottom border of the figure and drag it down a few rows. Your chart should look similar to the one shown below.

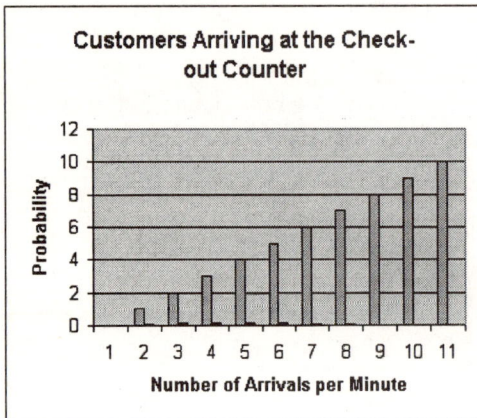

13. Correct the number scales displayed on the Y-axis and the X-axis. **Right click** on one of the vertical bars. Select **Source Data** from the shortcut menu that appears.

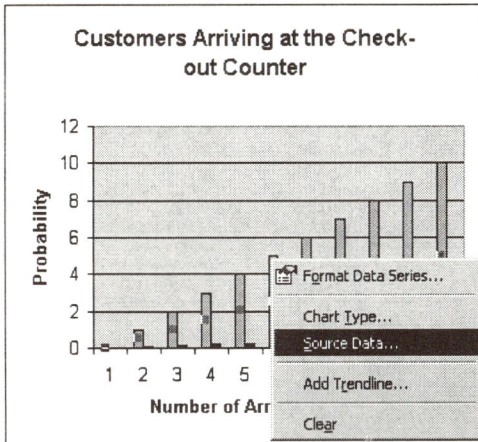

14. Click the **Series** tab at the top of the Source Data dialog box. In the Series window, select Series1 by clicking on it. Then click the **Remove** button below the Series window.

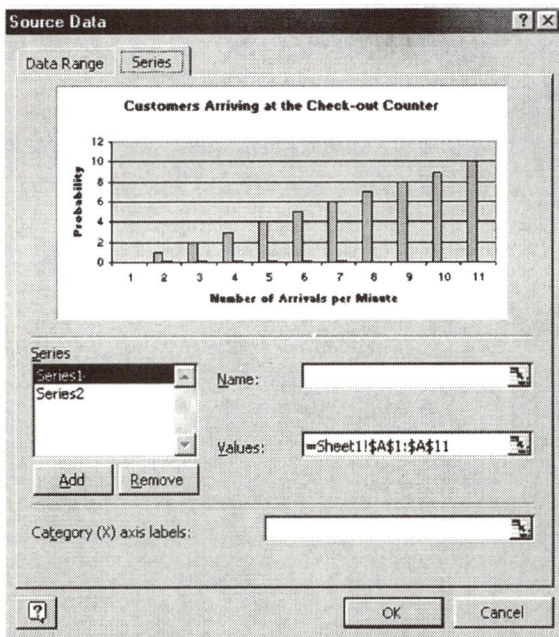

15.   The probabilities that should be displayed on the Y-axis are located in column B of
the worksheet. The entry in the Values window should be **=Sheet1!$B$1:$B$11**.
Revise the entry if necessary. First delete the information in the Values window.
Then click in the Values window. Next click on cell B1 of the worksheet and drag
down to cell B11. In the Values window you should now see
=Sheet1!$B$1:$B$11.

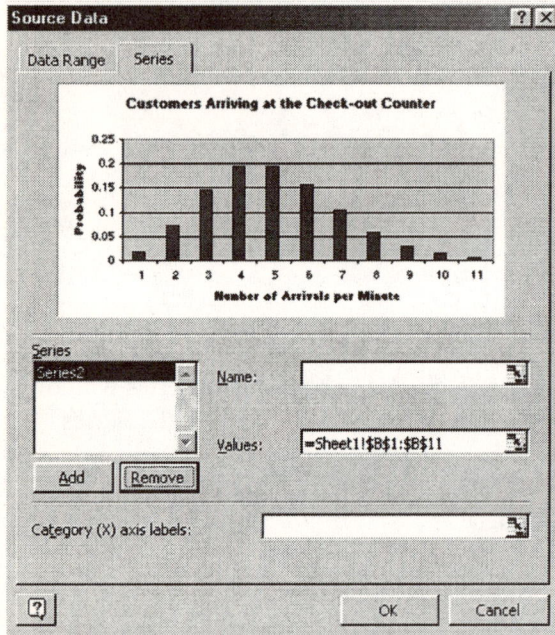

16. The numbers you want displayed on the X-axis are in column A of the worksheet. Click in the Category (X) axis labels window. Then click on cell A1 of the worksheet and drag down to cell A11. The entry in the Category (X) axis labels field should now read **=Sheet1!$A$1:$A$11**. Click **OK**.

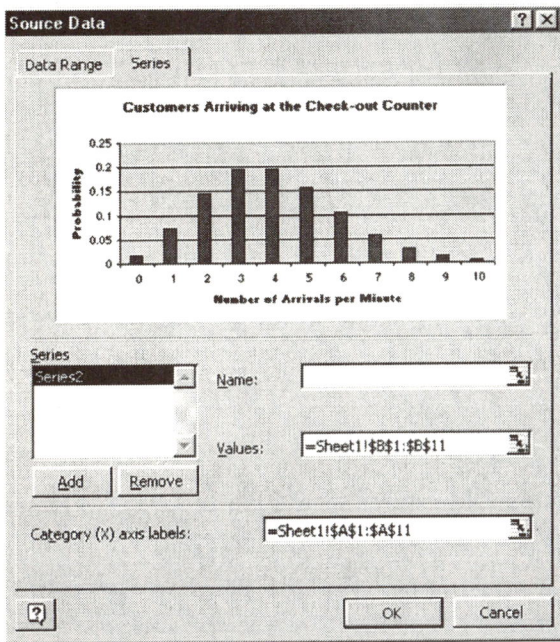

17. Remove the space between the vertical bars. **Right click** on one of the vertical bars. Select **Format Data Series** from the shortcut menu that appears.

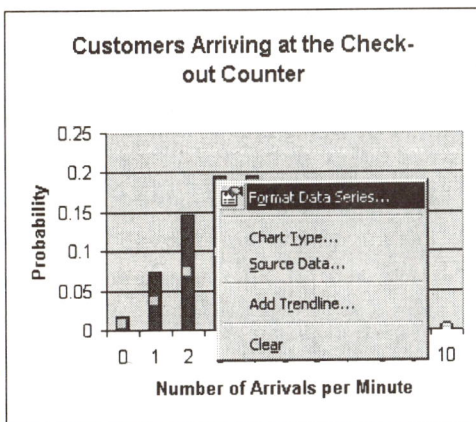

18. Click the **Options** tab at the top of the Format Data Series dialog box. Change the value in the Gap width box to 0. Click **OK**. Your histogram should look similar to the one shown below.

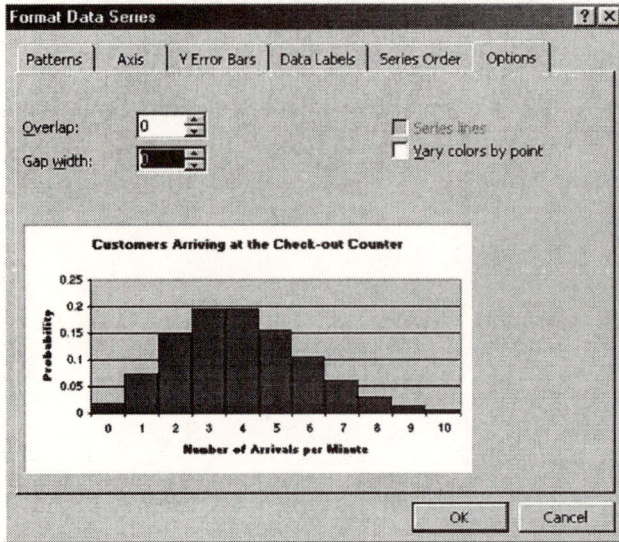

► Exercise 3 (pg. 195)	Generating a List of 20 Random Numbers with a Poisson distribution for $\mu = 4$

1. Open a new, blank Excel worksheet. At the top of the screen, click **Tools → Data Analysis**. Select **Random Number Generation** and click **OK**.

If Data Analysis does not appear as a choice in the Tools menu, you will need to load the Microsoft Excel Analysis ToolPak add-in. Follow the procedure in Section GS 8.1 before continuing.

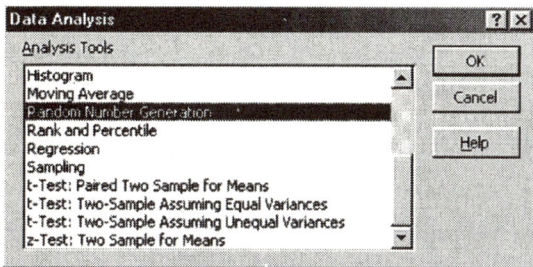

2. Complete the Random Number Generation dialog box as shown below. The **Number of Variables** indicates the number of columns of values that you want in the output table. **Number of Random Numbers** you want generated is 20. Select the **Poisson** distribution. **Lambda** is the expected value (μ), which is equal to 4. The output will be placed in the current worksheet with A1 as the left topmost cell. Click **OK**.

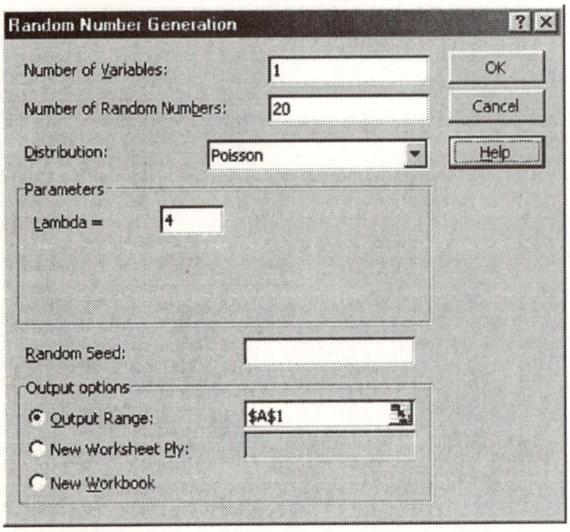

Random Number Generation		? X
Number of Variables:	1	OK
Number of Random Numbers:	20	Cancel
Distribution:	Poisson	Help
Parameters		
Lambda =	4	
Random Seed:		
Output options		
⦿ Output Range:	$A$1	
◯ New Worksheet Ply:		
◯ New Workbook		

The 20 numbers generated for this example are shown below. Because the numbers were generated randomly, it is not likely that your numbers will be exactly the same.

	A	B	C	D	E	F	G
1	3						
2	3						
3	4						
4	4						
5	5						
6	4						
7	5						
8	4						
9	5						
10	4						
11	3						
12	2						
13	4						
14	1						
15	6						
16	3						
17	7						
18	4						
19	3						
20	6						

Normal Probability Distributions

Section 5.2

▶ Example 2 (pg. 217)	Finding Area Under the Standard Normal Curve

1. Open a new Excel worksheet and click in cell **A1** to place the output there.

2. At the top of the screen, click **Insert → Function**.

3. Under Function category, select **Statistical**. Under Function name, select **NORMSDIST**. Click **OK**.

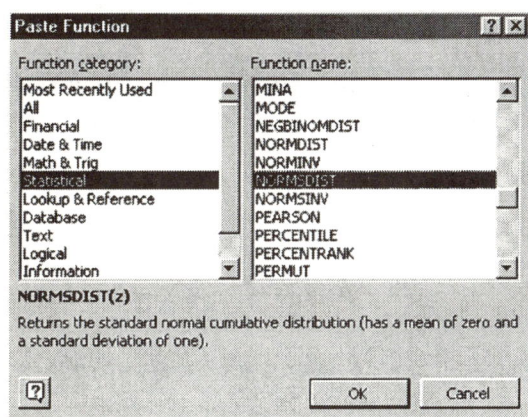

4. Complete the NORMSDIST dialog box as shown below. Click **OK**.

The output is displayed in cell A1 of the worksheet. The area under the standard normal curve to the left of z = -0.99 is 0.161087.

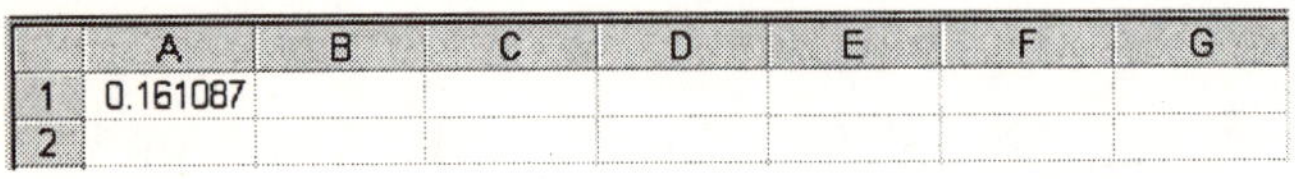

▶ Example 3 (pg. 217)   Finding Area Under the Standard Normal Curve

1. Open a new Excel worksheet. The area to the right of z = 1.06 is equal to 1 minus the area to the left of z = 1.06. In cell A1, enter **=1-** as shown below.

2. At the top of the screen, click **Insert → Function**.

3. Under Function category, select **Statistical**. Under Function name, select **NORMSDIST**. Click **OK**.

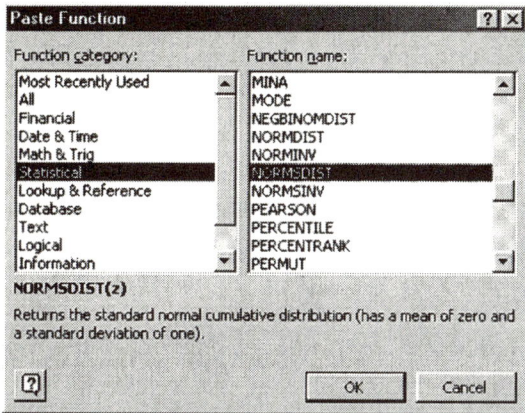

4. Complete the NORMSDIST dialog box as shown below. Click **OK**.

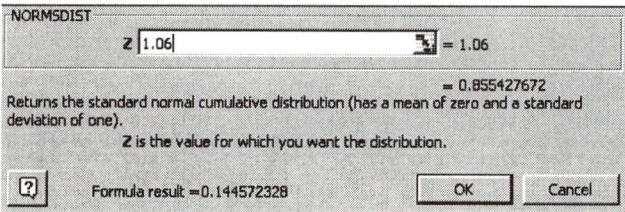

The output is displayed in cell A1 of the worksheet. The area under the standard normal curve to the right of z = 1.06 is 0.144572.

	A	B	C	D	E	F	G
1	0.144572						
2							

Section 5.3

| ► Example 3 (pg. 224) | Finding Normal Probabilities |

1. Open a new Excel worksheet and click in cell **A1** to place the output there.

2. At the top of the screen, click **Insert → Function**.

3. Under Function category, select **Statistical**. Under Function name, select **NORMDIST**. Click **OK**.

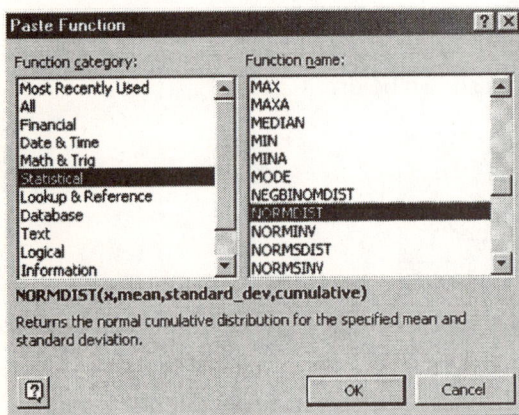

4. Complete the NORMDIST dialog box as shown below. Click **OK**.

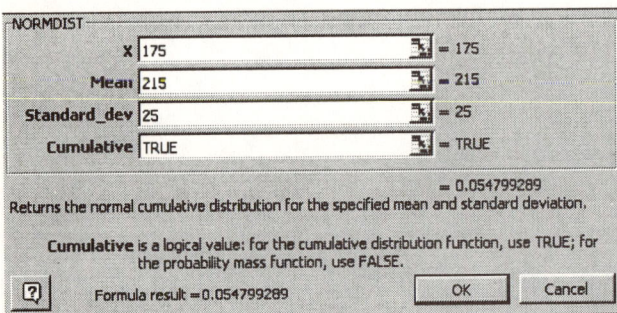

The output is displayed in cell A1 of the worksheet. The probability is equal to 0.054799.

	A	B	C	D	E	F	G
1	0.054799						
2							

◀

Technology Lab

▶ Exercise 1 (pg. 250)	Finding the Mean Age in the United States

1. Open a new, blank Excel worksheet. Enter **Class Midpoint** in cell A1. You will be entering the numbers displayed in the table on page 250 of your text. Begin by entering the first two midpoints, **2.5** and **7.5**, as shown below.

	A	B	C	D	E	F	G
1	Class Midpoint						
2	2.5						
3	7.5						

2. You will now fill column A with a series. Click in cell A2 and drag down to cell A3 so that both cells are highlighted.

	A	B	C	D	E	F	G
1	Class Midpoint						
2	2.5						
3	7.5						

3. Move the mouse pointer in cell A3 to the right lower corner of the cell so that the white plus sign turns into a black plus sign. The black plus sign is called the "fill handle." Click the left mouse button and drag the fill handle down to cell A21.

	A	B	C	D	E	F	G
1	Class Midpoint						
2	2.5						
3	7.5						
4	12.5						
5	17.5						
6	22.5						
7	27.5						
8	32.5						
9	37.5						
10	42.5						
11	47.5						
12	52.5						
13	57.5						
14	62.5						
15	67.5						
16	72.5						
17	77.5						
18	82.5						
19	87.5						
20	92.5						
21	97.5						

4. Enter **Relative Frequency** in cell B1. Then enter the proportion equivalents of the percentages in column B as shown below.

	A	B	C	D	E	F	G
1	Class Midpoint	Relative Frequency					
2	2.5	0.069					
3	7.5	0.071					
4	12.5	0.072					
5	17.5	0.072					
6	22.5	0.068					
7	27.5	0.064					
8	32.5	0.071					
9	37.5	0.08					
10	42.5	0.082					
11	47.5	0.073					
12	52.5	0.064					
13	57.5	0.049					
14	62.5	0.039					
15	67.5	0.034					
16	72.5	0.032					
17	77.5	0.027					
18	82.5	0.018					
19	87.5	0.01					
20	92.5	0.004					
21	97.5	0.001					

5. Click in cell C2 and enter a formula to multiply the midpoint by the relative frequency. The formula is **=A2*B2**. Press [**Enter**].

	A	B	C	D	E	F	G
1	Class Midpoint	Relative Frequency					
2	2.5	0.069	=A2*B2				

6. Copy the contents of cell C2 to cells C3 through C21.

	A	B	C	D	E	F	G
1	Class Midpoint	Relative Frequency					
2	2.5	0.069	0.1725				
3	7.5	0.071	0.5325				
4	12.5	0.072	0.9				
5	17.5	0.072	1.26				
6	22.5	0.068	1.53				
7	27.5	0.064	1.76				
8	32.5	0.071	2.3075				
9	37.5	0.08	3				
10	42.5	0.082	3.485				
11	47.5	0.073	3.4675				
12	52.5	0.064	3.36				
13	57.5	0.049	2.8175				
14	62.5	0.039	2.4375				
15	67.5	0.034	2.295				
16	72.5	0.032	2.32				
17	77.5	0.027	2.0925				
18	82.5	0.018	1.485				
19	87.5	0.01	0.875				
20	92.5	0.004	0.37				
21	97.5	0.001	0.0975				

7. The weighted mean is equal to the sum of the products in column C. (Refer to Section 2.3 in your text.) Click in cell **C22** of the worksheet to place the sum there. Click the AutoSum button near the top of the screen. It looks like this: Σ. The range of numbers to be included in the sum is displayed in cell C22. You should see **=SUM(C2:C21)**. Make any necessary corrections. Then press [**Enter**].

9	37.5	0.08	3				
10	42.5	0.082	3.485				
11	47.5	0.073	3.4675				
12	52.5	0.064	3.36				
13	57.5	0.049	2.8175				
14	62.5	0.039	2.4375				
15	67.5	0.034	2.295				
16	72.5	0.032	2.32				
17	77.5	0.027	2.0925				
18	82.5	0.018	1.485				
19	87.5	0.01	0.875				
20	92.5	0.004	0.37				
21	97.5	0.001	0.0975				
22			=SUM(C2:C21)				

8. The mean, displayed in cell C22, is 36.565 years. Save this worksheet so that you can use it again for Exercise 5, page 250.

◀

| ▶ Exercise 2 (pg. 250) | Finding the Mean of the Set of Sample Means |

1. Open worksheet "Tech5" in the Chapter 5 folder.

2. Click in cell **A38** at the bottom of the column of numbers to place the mean in that cell.

	A	B	C	D	E	F	G
34	32.88						
35	37.33						
36	31.27						
37	35.8						
38							

3. At the top of the screen, click **Insert → Function**.

4. Under Function category, select **Statistical**. Under Function name, select **AVERAGE**. Click **OK**.

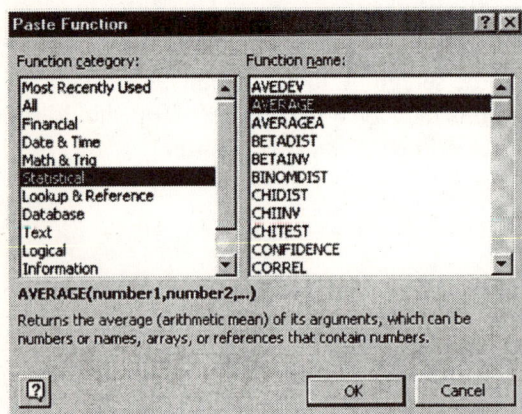

Paste Function   ? ✕

Function category:	Function name:
Most Recently Used	AVEDEV
All	AVERAGE
Financial	AVERAGEA
Date & Time	BETADIST
Math & Trig	BETAINV
Statistical	BINOMDIST
Lookup & Reference	CHIDIST
Database	CHIINV
Text	CHITEST
Logical	CONFIDENCE
Information	CORREL

AVERAGE(number1,number2,...)

Returns the average (arithmetic mean) of its arguments, which can be numbers or names, arrays, or references that contain numbers.

[?]   OK   Cancel

5. The range that will be included in the average is shown in the Number 1 window. Check to be sure that it is accurate. It should read **A2:A37**. Make any necessary corrections. Then click **OK**.

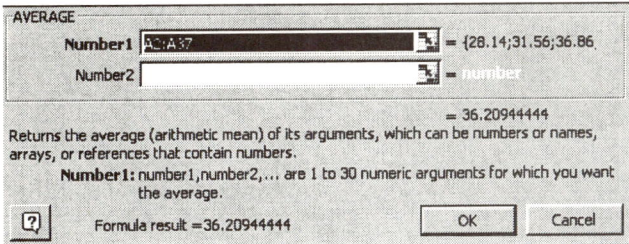

The mean of the sample means is displayed in cell A38 of the worksheet. The mean is equal to 36.20944.

◀

▶ Exercise 5 (pg. 250)	Finding the Standard Deviation of Ages in the United States

1. Open the worksheet that you prepared for Exercise 1, page 250.

	A	B	C	D	E	F	G
1	Class Midpoint	Relative Frequency					
2	2.5	0.069	0.1725				
3	7.5	0.071	0.5325				
4	12.5	0.072	0.9				
5	17.5	0.072	1.26				
6	22.5	0.068	1.53				
7	27.5	0.064	1.76				

2. To find the standard deviation of the population of ages, you will first calculate squared deviation scores. Click in cell D1 and enter the label, **Sqd Dev Score**.

	A	B	C	D	E	F	G
1	Class Midpoint	Relative Frequency		Sqd Dev Score			
2	2.5	0.069	0.1725				

3. Click in cell **D2** and enter the formula =(A2-36.565)^2 and press [**Enter**].

	A	B	C	D	E	F	G
1	Class Midpoint	Relative Frequency		Sqd Dev Score			
2	2.5	0.069	0.1725	=(A2-36.565)^2			

4. Copy the formula in cell D2 to cells D3 through D21.

	A	B	C	D	E	F	G
1	Class Midpoint	Relative Frequency		Sqd Dev Score			
2	2.5	0.069	0.1725	1160.424			
3	7.5	0.071	0.5325	844.7742			
4	12.5	0.072	0.9	579.1242			
5	17.5	0.072	1.26	363.4742			
6	22.5	0.068	1.53	197.8242			
7	27.5	0.064	1.76	82.17423			
8	32.5	0.071	2.3075	16.52423			
9	37.5	0.08	3	0.874225			
10	42.5	0.082	3.485	35.22423			
11	47.5	0.073	3.4675	119.5742			
12	52.5	0.064	3.36	253.9242			
13	57.5	0.049	2.8175	438.2742			
14	62.5	0.039	2.4375	672.6242			
15	67.5	0.034	2.295	956.9742			
16	72.5	0.032	2.32	1291.324			
17	77.5	0.027	2.0925	1675.674			
18	82.5	0.018	1.485	2110.024			
19	87.5	0.01	0.875	2594.374			
20	92.5	0.004	0.37	3128.724			
21	97.5	0.001	0.0975	3713.074			

5. Each of the squared deviations will be weighted by its relative frequency. Click in cell E2. Enter the formula =**D2*B2** and press [**Enter**].

	A	B	C	D	E	F	G
1	Class Midpoint	Relative Frequency		Sqd Dev Score			
2	2.5	0.069	0.1725	1160.424	=D2*B2		

6. Copy the formula in cell E2 to cells E3 through E21.

	A	B	C	D	E	F	G
1	Class Midpoint	Relative Frequency		Sqd Dev Score			
2	2.5	0.069	0.1725	1160.424	80.06927		
3	7.5	0.071	0.5325	844.7742	59.97897		
4	12.5	0.072	0.9	579.1242	41.69694		
5	17.5	0.072	1.26	363.4742	26.17014		
6	22.5	0.068	1.53	197.8242	13.45205		
7	27.5	0.064	1.76	82.17423	5.25915		
8	32.5	0.071	2.3075	16.52423	1.17322		
9	37.5	0.08	3	0.874225	0.069938		
10	42.5	0.082	3.485	35.22423	2.888386		
11	47.5	0.073	3.4675	119.5742	8.728918		
12	52.5	0.064	3.36	253.9242	16.25115		
13	57.5	0.049	2.8175	438.2742	21.47544		
14	62.5	0.039	2.4375	672.6242	26.23234		
15	67.5	0.034	2.295	956.9742	32.53712		
16	72.5	0.032	2.32	1291.324	41.32238		
17	77.5	0.027	2.0925	1675.674	45.2432		
18	82.5	0.018	1.485	2110.024	37.98044		
19	87.5	0.01	0.875	2594.374	25.94374		
20	92.5	0.004	0.37	3128.724	12.5149		
21	97.5	0.001	0.0975	3713.074	3.713074		

7. You are working with a relative frequency distribution of midpoints. For this type of distribution, the variance is equal to the sum of the squared deviation scores. Click in cell **E22** to place the sum of the squared deviation scores there. Click the AutoSum button near the top of the screen. It looks like this: Σ. The range of numbers to be included in the sum is displayed in cell E22. You should see **=SUM(E2:E21)**. Make any necessary corrections. Then press [**Enter**].

18	82.5	0.018	1.485	2110.024	37.98044	
19	87.5	0.01	0.875	2594.374	25.94374	
20	92.5	0.004	0.37	3128.724	12.5149	
21	97.5	0.001	0.0975	3713.074	3.713074	
22			36.565		=SUM(E2:E21)	

8. Click in cell E23 to place the standard deviation there. The standard deviation is the square root of the variance. In cell E23, enter the formula **=sqrt(E22)** and press [**Enter**].

18	82.5	0.018	1.485	2110.024	37.98044	
19	87.5	0.01	0.875	2594.374	25.94374	
20	92.5	0.004	0.37	3128.724	12.5149	
21	97.5	0.001	0.0975	3713.074	3.713074	
22			36.565		502.7008	
23					=sqrt(E22)	

9. The standard deviation of ages in the United States is approximately equal to 22.42. For future reference purposes, you may want to add labels for the mean, variance, and standard deviation as shown below.

18		82.5	0.018	1.485	2110.024	37.98044		
19		87.5	0.01	0.875	2594.374	25.94374		
20		92.5	0.004	0.37	3128.724	12.5149		
21		97.5	0.001	0.0975	3713.074	3.713074		
22			Mean =	36.565	Var =	502.7008		
23					St Dev =	22.42099		

◄

| ► Exercise 6 (pg. 250) | Finding the Standard Deviation of the Set of Sample Means |

1. Open worksheet "Tech5" in the Chapter 5 folder. This is the same data set that you used for Exercise 2, page 250. If you placed the mean (36.20944) in cell A38 for Exercise 2, page 250, you should delete it now.

	A	B	C	D	E	F	G
1	mean ages						
2	28.14						
3	31.56						
4	36.86						

2. At the top of the screen, click **Tools** → **Data Analysis**.

If Data Analysis does not appear as a choice in the Tools menu, you will need to load the Microsoft Excel Analysis ToolPak add-in. Follow the procedure in Section GS 8.1 before continuing.

3. Select **Descriptive Statistics** and click **OK**.

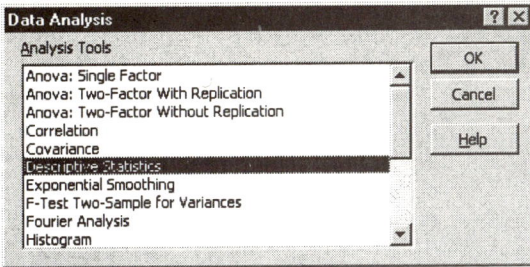

4. Complete the Descriptive Statistics dialog box as shown below. Click **OK**.

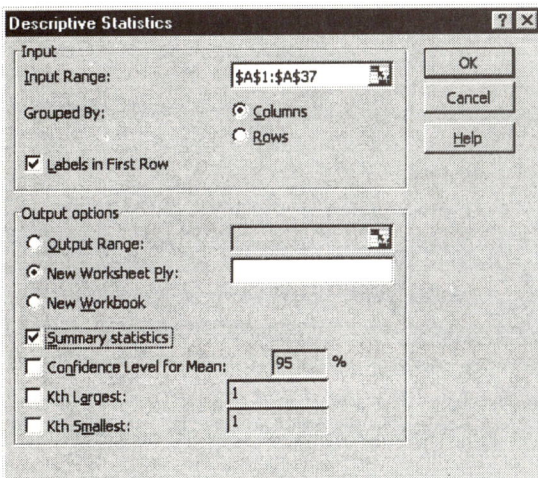

You will want to make column A wider so that you can read all the labels in the output table. The standard deviation is equal to 3.551804.

	A	B
1	*mean ages*	
2		
3	Mean	36.20944
4	Standard Error	0.591967
5	Median	36.155
6	Mode	#N/A
7	Standard Deviation	3.551804
8	Sample Variance	12.61531
9	Kurtosis	0.343186
10	Skewness	0.207283
11	Range	16.58
12	Minimum	28.14
13	Maximum	44.72
14	Sum	1303.54
15	Count	36

Confidence Intervals

Section 6.1

| ► Example 4 (pg. 274) | Constructing a Confidence Interval |

If the PHStat2 add-in has not been loaded, you will need to load it before continuing. Follow the instructions in Section GS 8.2.

1. Open the "Sentence" worksheet in the Chapter 6 folder.

2. At the top of the screen, select **PHStat → Confidence Intervals → Estimate for the Mean, sigma known**.

3. Complete the Estimate for the Mean dialog box as shown below. Click **OK**.

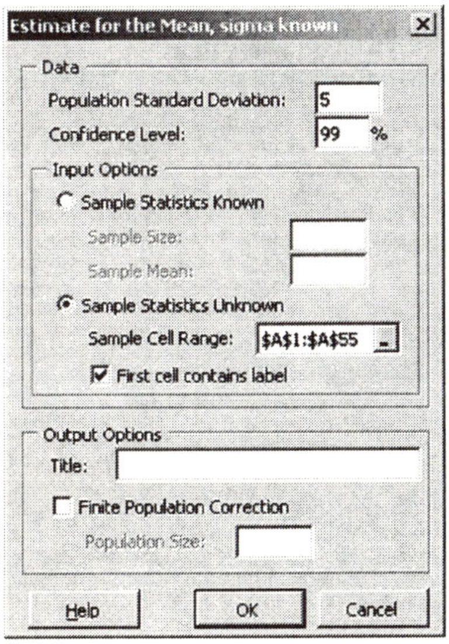

The output is displayed in a new worksheet. The lower limit of the confidence interval is 10.7 and the upper limit is 14.2.

	A	B
1	Confidence Interval Estimate for the Mean	
2		
3	Data	
4	Population Standard Deviation	5
5	Sample Mean	12.42592593
6	Sample Size	54
7	Confidence Level	99%
8		
9	Intermediate Calculations	
10	Standard Error of the Mean	0.680413817
11	Z Value	-2.57583451
12	Interval Half Width	1.752633395
13		
14	Confidence Interval	
15	Interval Lower Limit	10.67329253
16	Interval Upper Limit	14.17855932

◀

▶ Example 5 (pg. 275)	Constructing a Confidence Interval, σ Known

If the PHStat2 add-in has not been loaded, you will need to load it before continuing. Follow the instructions in Section GS 8.2.

1. Open a new, blank Excel worksheet. At the top of the screen, select **PHStat** → **Confidence Intervals** → **Estimate for the Mean, sigma known**.

2.   Complete the Estimate for the Mean dialog box as shown below. Click **OK**.

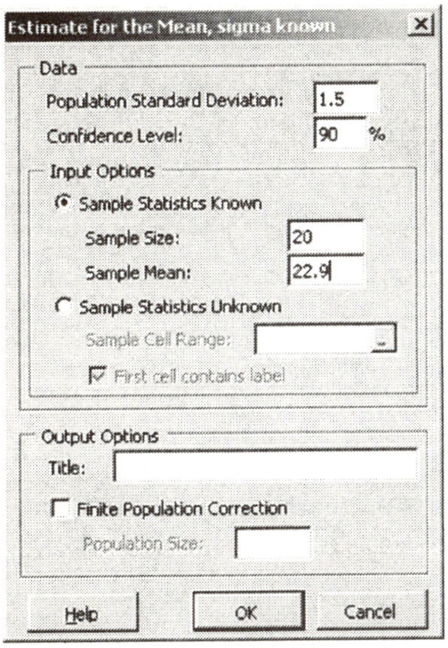

Estimate for the Mean, sigma known   ☒

Data

Population Standard Deviation: 1.5

Confidence Level: 90 %

Input Options

⦿ Sample Statistics Known

Sample Size: 20

Sample Mean: 22.9

○ Sample Statistics Unknown

Sample Cell Range:

☑ First cell contains label

Output Options

Title:

☐ Finite Population Correction

Population Size:

| Help | OK | Cancel |

Your output should look the same as the output displayed below. The lower limit of the 90% confidence interval is 22.3 and the upper limit is 23.5.

	A	B
1	**Confidence Interval Estimate for the Mean**	
2		
3	**Data**	
4	**Population Standard Deviation**	1.5
5	**Sample Mean**	22.9
6	**Sample Size**	20
7	**Confidence Level**	90%
8		
9	Intermediate Calculations	
10	Standard Error of the Mean	0.335410197
11	Z Value	-1.644853
12	Interval Half Width	0.551700468
13		
14	Confidence Interval	
15	**Interval Lower Limit**	22.34829953
16	**Interval Upper Limit**	23.45170047

Section 6.2

> ▶ Example 2 (pg. 286) | Constructing a confidence interval, σ unknown

If the PHStat2 add-in has not been loaded, you will need to load it before continuing. Follow the instructions in Section GS 8.2.

1. Open a new, blank Excel worksheet. At the top of the screen, select **PHStat →** **Confidence Intervals → Estimate for the Mean, sigma unknown**.

2. Complete the Estimate for the Mean dialog box as shown below. Click **OK**.

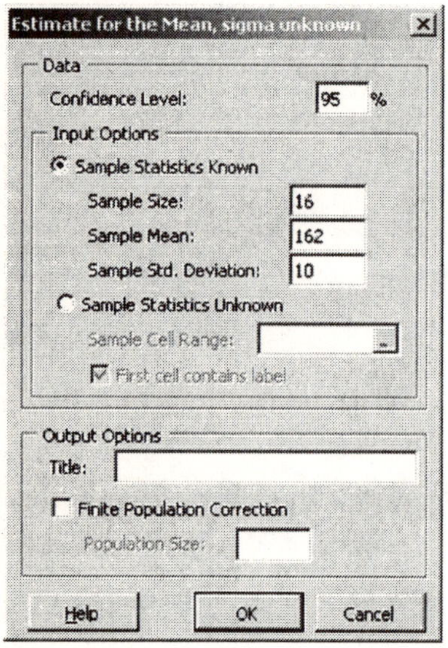

Estimate for the Mean, sigma unknown ☒

- Data
  - Confidence Level: `95` %
  - Input Options
    - ⦿ Sample Statistics Known
      - Sample Size: `16`
      - Sample Mean: `162`
      - Sample Std. Deviation: `10`
    - ○ Sample Statistics Unknown
      - Sample Cell Range: [   ] [..]
      - ☑ First cell contains label

- Output Options
  - Title: [   ]
  - ☐ Finite Population Correction
    - Population Size: [   ]

[Help]   [OK]   [Cancel]

The output is displayed in a new worksheet. The lower limit of the 95% confidence interval is 156.67 and the upper limit is 167.33.

	A	B
1	**Confidence Interval Estimate for the Mean**	
2		
3	Data	
4	**Sample Standard Deviation**	10
5	**Sample Mean**	162
6	**Sample Size**	16
7	**Confidence Level**	95%
8		
9	Intermediate Calculations	
10	Standard Error of the Mean	2.5
11	Degrees of Freedom	15
12	*t* Value	2.131450856
13	Interval Half Width	5.32862714
14		
15	**Confidence Interval**	
16	**Interval Lower Limit**	156.67
17	**Interval Upper Limit**	167.33

◄

Section 6.3

▶ **Example 2 (pg. 295)**   Constructing a confidence interval for p

If the PHStat2 add-in has not been loaded, you will need to load it before continuing. Follow the instructions in Section GS 8.2.

1. Open a new, blank Excel worksheet. At the top of the screen, select **PHStat → Confidence Intervals → Estimate for the Proportion**.

2. Complete the Estimate for the Proportion dialog box as shown at the top of the next page. Click **OK**.

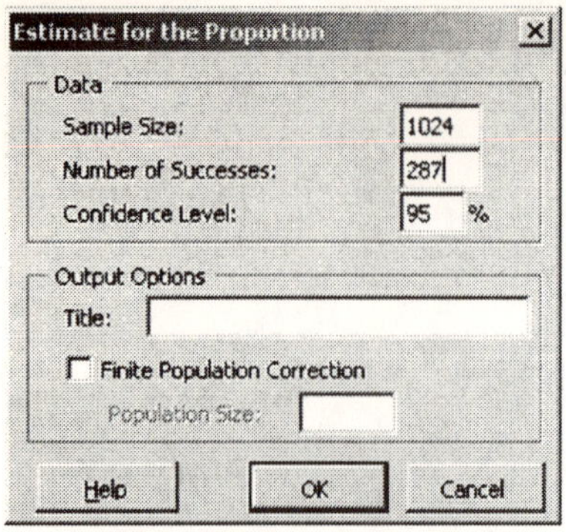

Your output should look like the output displayed below. The lower limit of the 95% confidence interval is 0.253 and the upper limit is 0.308.

	A	B
1	Confidence Interval Estimate for the Mean	
2		
3	Data	
4	Sample Size	1024
5	Number of Successes	287
6	Confidence Level	95%
7		
8	Intermediate Calculations	
9	Sample Proportion	0.280273438
10	Z Value	-1.95996108
11	Standard Error of the Proportion	0.014035399
12	Interval Half Width	0.027508835
13		
14	Confidence Interval	
15	Interval Lower Limit	0.252764602
16	Interval Upper Limit	0.307782273

◀

Technology Lab

▶ **Exercise 1 (pg. 301)**   Finding a 95% Confidence Interval for p

If the PHStat2 add-in has not been loaded, you will need to load it before continuing. Follow the instructions in Section GS 8.2.

1.   Open a new, blank Excel worksheet. At the top of the screen, select **PHStat** →
     Confidence Intervals → **Estimate for the Proportion**.

2.   Complete the Estimate for the Proportion dialog box as shown below. Click **OK**.

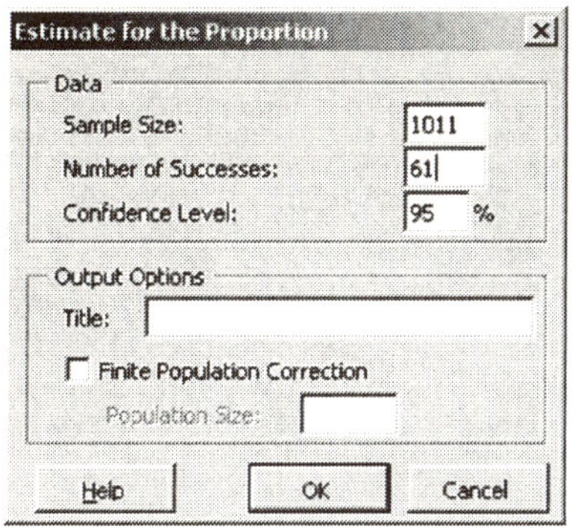

Your output should look the same as the output displayed below. The lower limit of the 95% confidence interval is 0.046 and the upper limit is 0.075.

	A	B
1	Confidence Interval Estimate for the Mean	
2		
3	Data	
4	Sample Size	1011
5	Number of Successes	61
6	Confidence Level	95%
7		
8	Intermediate Calculations	
9	Sample Proportion	0.060336301
10	Z Value	-1.95996108
11	Standard Error of the Proportion	0.007488589
12	Interval Half Width	0.014677343
13		
14	Confidence Interval	
15	Interval Lower Limit	0.045658958
16	Interval Upper Limit	0.075013643

◀

▶ Exercise 3 (pg. 301)	Finding a 95% confidence interval for p

If the PHStat2 add-in has not been loaded, you will need to load it before continuing. Follow the instructions in Section GS 8.2.

1.   Open a new, blank Excel worksheet. At the top of the screen, select **PHStat →
     Confidence Intervals → Estimate for the Proportion**.

2. Four percent of the sample size, 1011, is 40. Use this number to complete the
    Estimate for the Proportion dialog box as shown below. Click **OK**.

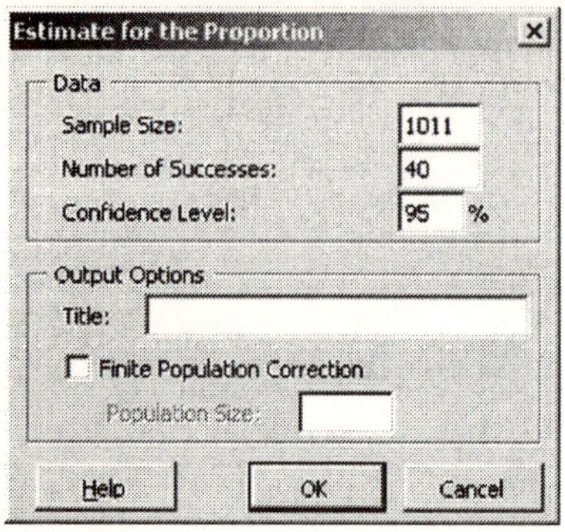

Estimate for the Proportion	✕
Data	
Sample Size:	1011
Number of Successes:	40
Confidence Level:	95 %
Output Options	
Title:	
☐ Finite Population Correction	
Population Size:	
Help	OK	Cancel

Your output should look the same as the output displayed below. The upper limit of the
95% confidence interval is 0.028 and the upper limit is 0.052.

	A	B
1	**Confidence Interval Estimate for the Mean**	
2		
3	**Data**	
4	**Sample Size**	1011
5	**Number of Successes**	40
6	**Confidence Level**	95%
7		
8	Intermediate Calculations	
9	Sample Proportion	0.039564787
10	Z Value	-1.95996108
11	Standard Error of the Proportion	0.00613074
12	Interval Half Width	0.012016011
13		
14	**Confidence Interval**	
15	**Interval Lower Limit**	0.027548776
16	**Interval Upper Limit**	0.051580799

◄

► Exercise 4 (pg. 301)	Simulating a Most Admired Poll

1. Open a new, blank Excel worksheet. Enter the labels shown at the top of the next
    page for displaying the output of five simulations.

	A	B	C	D	E	F	G
1	Time 1						
2	Time 2						
3	Time 3						
4	Time 4						
5	Time 5						

2. At the top of the screen, select **Tools** → **Data Analysis**. Select **Random Number Generation**. Click **OK**.

If Data Analysis does not appear as a choice in the Tools menu, you will need to load the Microsoft Excel Analysis ToolPak add-in. Follow the procedure in Section GS 8.1 before continuing.

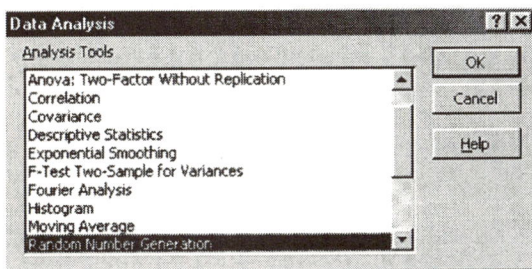

3. Complete the Random Number Generation dialog box as shown below. Click **OK**.

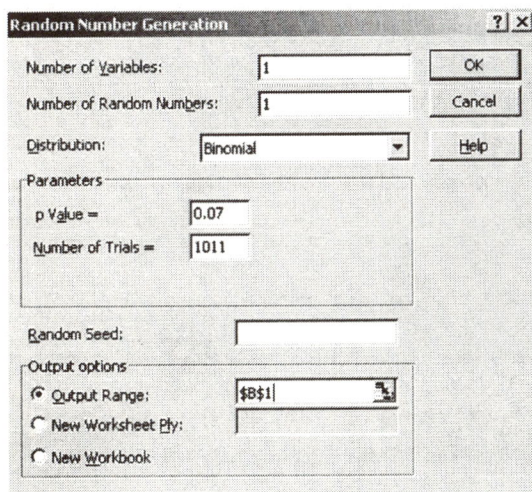

For Time 1, the number of persons (out of 1011) selecting Oprah Winfrey is 72.

	A	B	C	D	E	F	G
1	Time 1	72					
2	Time 2						
3	Time 3						
4	Time 4						
5	Time 5						

4. Begin Time 2 the same way that you did Time 1. At the top of the screen, click **Tools → Data Analysis**. Select **Random Number Generation**. Click **OK**.

5. Complete the Random Number Generation dialog box in the same way as you did for Time 1 except change the location of the output to cell B2 as shown below. Click **OK**.

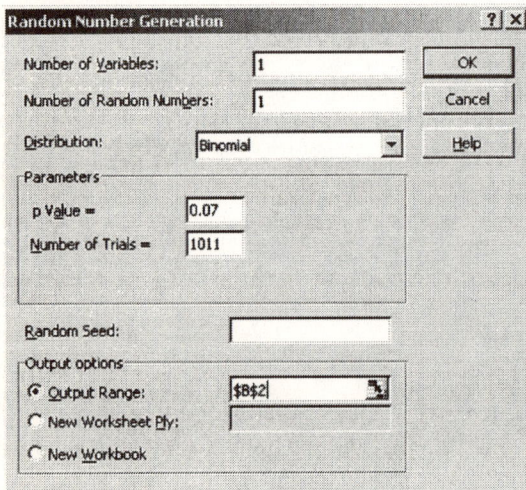

Random Number Generation

Number of Variables: 1

Number of Random Numbers: 1

Distribution: Binomial

Parameters

p Value = 0.07

Number of Trials = 1011

Random Seed:

Output options

⊙ Output Range: $B$2

○ New Worksheet Ply:

○ New Workbook

OK   Cancel   Help

6. Repeat this procedure until you have carried out the simulation five times. Your output should be similar to the output displayed below. Because the numbers were generated randomly, however, it is not likely that your numbers will be exactly the same as these.

	A	B	C	D	E	F	G
1	Time 1	72					
2	Time 2	66					
3	Time 3	63					
4	Time 4	80					
5	Time 5	69					

Hypothesis Testing with One Sample

Section 7.2

► Example 6 (pg. 339)	Using a Technology Tool to Find a P-value

If the PHStat add-in has not been loaded, you will need to load it before continuing. Follow the instructions in Section GS 8.2.

1. First open a new Excel worksheet. Then, at the top of the screen, select **PHStat →** **One-Sample Tests → Z Test for the Mean, sigma known**.

2. Complete the Z Test for the Mean dialog box as shown below. Click **OK**.

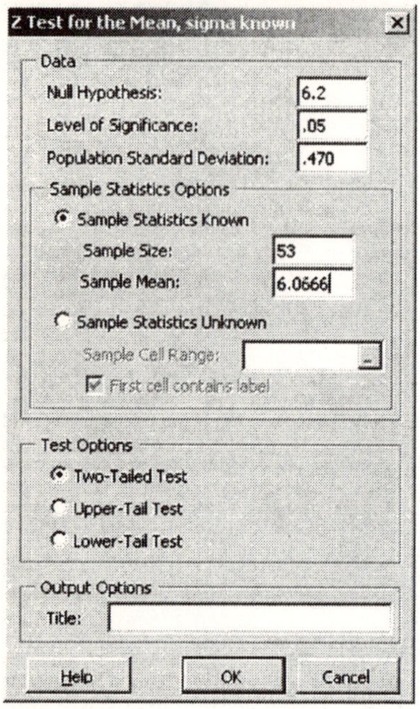

Your output should look like the output displayed below.

	A	B
1	Z Test of Hypothesis for the Mean	
2		
3	Data	
4	Null Hypothesis $\mu=$	6.2
5	Level of Significance	0.05
6	Population Standard Deviation	0.47
7	Sample Size	53
8	Sample Mean	6.0666
9		
10	Intermediate Calculations	
11	Standard Error of the Mean	0.064559465
12	Z Test Statistic	-2.066312041
13		
14	Two-Tailed Test	
15	Lower Critical Value	-1.959961082
16	Upper Critical Value	1.959961082
17	p-Value	0.038798892
18	Reject the null hypothesis	

▶ Example 9 (pg. 342)	Testing μ with a Large Sample

If the PHStat add-in has not been loaded, you will need to load it before continuing. Follow the instructions in Section GS 8.2.

3. First open a new Excel worksheet. Then, at the top of the screen, select **PHStat →
    One-Sample Tests → Z Test for the Mean, sigma known**.

4. Complete the Z Test for the Mean dialog box as shown below. Click **OK**.

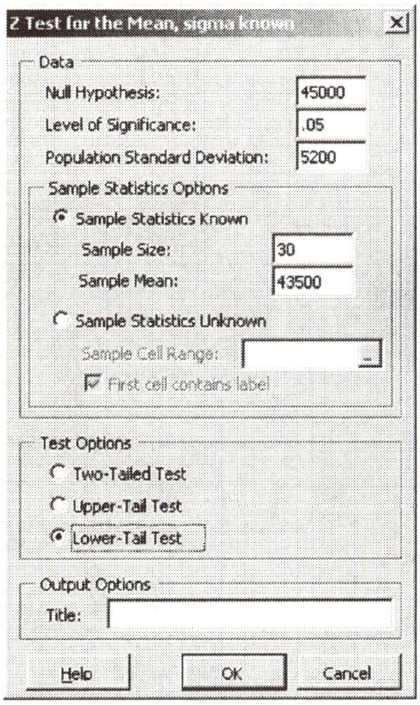

Your output should look like the output displayed below.

	A	B
1	Z Test of Hypothesis for the Mean	
2		
3	Data	
4	Null Hypothesis μ=	45000
5	Level of Significance	0.05
6	Population Standard Deviation	5200
7	Sample Size	30
8	Sample Mean	43500
9		
10	Intermediate Calculations	
11	Standard Error of the Mean	949.3857663
12	Z Test Statistic	-1.579968916
13		
14	Lower-Tail Test	
15	Lower Critical Value	-1.644853
16	p-Value	0.057056996
17	Do not reject the null hypothesis	

◄

> ▶ **Exercise 35 (pg. 347)**   **Testing That the Mean Time to Quit Smoking Is 15 Years**

If the PHStat add-in has not been loaded, you will need to load it before continuing. Follow the instructions in Section GS 8.2.

1. Open the "Ex7_2-35" worksheet in the Chapter 7 folder.

2. Your textbook indicates that the standard deviation of a sample may be used in the z-test formula instead of a known population standard deviation if the sample size is 30 or greater. You will use Excel's Descriptive Statistics tool to obtain the sample standard deviation. At the top of the screen, click **Tools → Data Analysis**. Select **Descriptive Statistics**. Click **OK**.

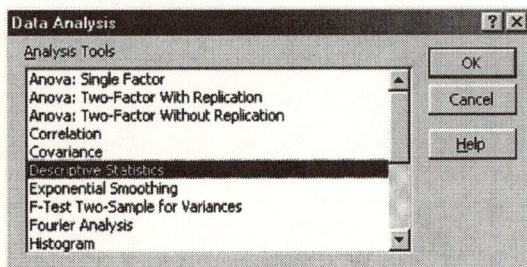

3. Complete the Descriptive Statistics dialog box as shown below. Click **OK**.

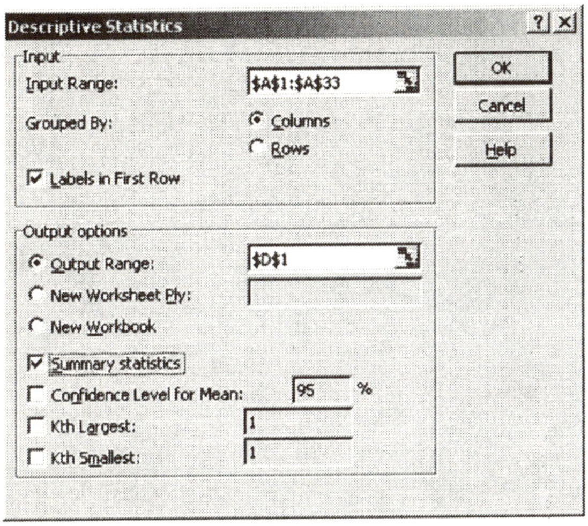

4. You will want to adjust the width of column D so that you can read all the labels.
    The standard deviation of the sample is 4.2876, the mean is 14.8344, and the count
    is 32. You will use this information when you carry out the statistical test. At the
    top of the screen, select **PHStat → One-Sample Tests → Z Test for the Mean,
    sigma known**.

	A	B	C	D	E
1	years to quit smoking			*years to quit smoking*	
2	15.7				
3	13.2			Mean	14.83438
4	22.6			Standard Error	0.75795
5	13			Median	14.65
6	10.7			Mode	13.2
7	18.1			Standard Deviation	4.287612
8	14.7			Sample Variance	18.38362
9	7			Kurtosis	-0.57155
10	17.3			Skewness	0.042962
11	7.5			Range	15.6
12	21.8			Minimum	7
13	12.3			Maximum	22.6
14	19.8			Sum	474.7
15	13.8			Count	32

5.   Complete the Z Test for the Mean dialog box as shown below. Click **OK**.

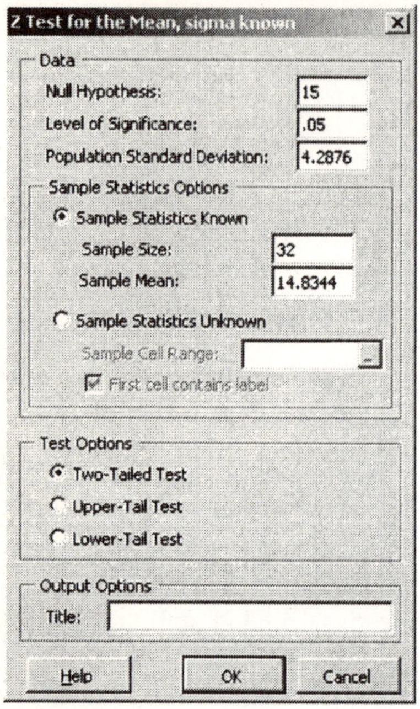

Z Test for the Mean, sigma known	✕
Data	
Null Hypothesis:	15
Level of Significance:	.05
Population Standard Deviation:	4.2876
Sample Statistics Options	
⦿ Sample Statistics Known	
Sample Size:	32
Sample Mean:	14.8344
○ Sample Statistics Unknown	
Sample Cell Range:	
☑ First cell contains label	
Test Options	
⦿ Two-Tailed Test	
○ Upper-Tail Test	
○ Lower-Tail Test	
Output Options	
Title:	
Help	OK	Cancel

Your output should look like the output displayed below.

	A	B
1	**Z Test of Hypothesis for the Mean**	
2		
3	**Data**	
4	**Null Hypothesis** $\mu=$	15
5	**Level of Significance**	0.05
6	**Population Standard Deviation**	4.2876
7	**Sample Size**	32
8	**Sample Mean**	14.8344
9		
10	Intermediate Calculations	
11	Standard Error of the Mean	0.757947759
12	Z Test Statistic	-0.218484715
13		
14	**Two-Tailed Test**	
15	**Lower Critical Value**	-1.959961082
16	**Upper Critical Value**	1.959961082
17	p-**Value**	0.827051536
18	**Do not reject the null hypothesis**	

Section 7.3

▶ Example 4 (pg. 353)	Testing μ with a Small Sample

If the PHStat add-in has not been loaded, you will need to load it before continuing. Follow the instructions in Section GS 8.2

1. First open a new Excel worksheet. Then, at the top of the screen, select **PHStat →
One-Sample Tests → t Test for the Mean, sigma unknown**.

2. Complete the t Test for the Mean dialog box as shown below. Click **OK**.

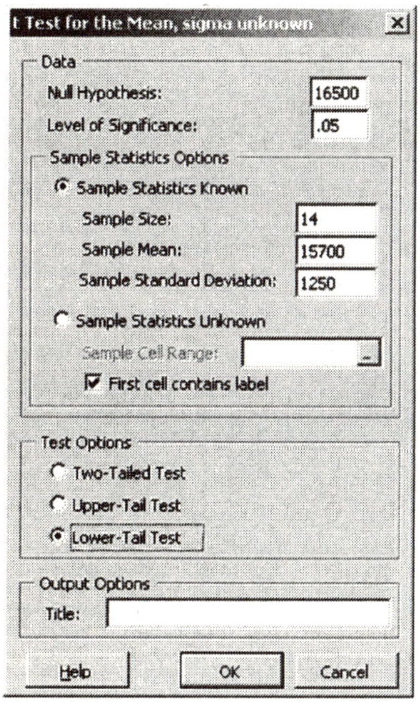

t Test for the Mean, sigma unknown

Data
Null Hypothesis: 16500
Level of Significance: .05

Sample Statistics Options
⊙ Sample Statistics Known
  Sample Size: 14
  Sample Mean: 15700
  Sample Standard Deviation: 1250

○ Sample Statistics Unknown
  Sample Cell Range:
  ☑ First cell contains label

Test Options
○ Two-Tailed Test
○ Upper-Tail Test
⊙ Lower-Tail Test

Output Options
Title:

Help   OK   Cancel

Your output should look like the output shown below.

	A	B	C	D	E
1	t Test for Hypothesis of the Mean				
2					
3	Data				
4	Null Hypothesis μ=	16500			
5	Level of Significance	0.05			
6	Sample Size	14			
7	Sample Mean	15700			
8	Sample Standard Deviation	1250			
9					
10	Intermediate Calculations				
11	Standard Error of the Mean	334.0765524			
12	Degrees of Freedom	13			
13	t Test Statistic	-2.394660728			
14					
15	Lower-Tail Test				
16	Lower Critical Value	-1.770931704			
17	p-Value	0.016203608			
18	Reject the null hypothesis				
19					
20				Calculations Area	
21				For one-tailed tests:	
22				TDIST value	0.016204
23				1-TDIST value	0.983796

◀

▶ Example 6 (pg. 355)	Using P-Values with a t-Test

If the PHStat add-in has not been loaded, you will need to load it before continuing. Follow the instructions in Section GS 8.2.

1. First open a new Excel worksheet. Then, at the top of the screen, select **PHStat →
   One-Sample Tests → t Test for the Mean, sigma unknown**.

2.   Complete the t Test for the Mean dialog box as shown below. Click **OK**.

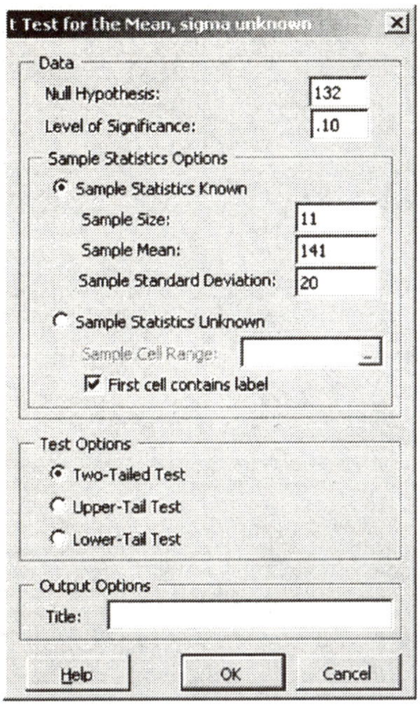

Your output should look like the output displayed below.

	A	B
1	t Test for Hypothesis of the Mean	
2		
3	Data	
4	Null Hypothesis μ=	132
5	Level of Significance	0.1
6	Sample Size	11
7	Sample Mean	141
8	Sample Standard Deviation	20
9		
10	Intermediate Calculations	
11	Standard Error of the Mean	6.030226892
12	Degrees of Freedom	10
13	t Test Statistic	1.492481156
14		
15	Two-Tailed Test	
16	Lower Critical Value	-1.812461505
17	Upper Critical Value	1.812461505
18	p-Value	0.166433529
19	Do not reject the null hypothesis	

> ► Exercise 25 (pg. 357)

Testing the Claim That the Mean Recycled Waste Is More Than 1 Lb. Per Day

If the PHStat add-in has not been loaded, you will need to load it before continuing. Follow the instructions in Section GS 8.2.

1. First open a new Excel worksheet. Then, at the top of the screen, select **PHStat →
   One-Sample Tests → t Test for the Mean, sigma unknown**.

2. Complete the t Test for the Mean dialog box as shown below. Click **OK**.

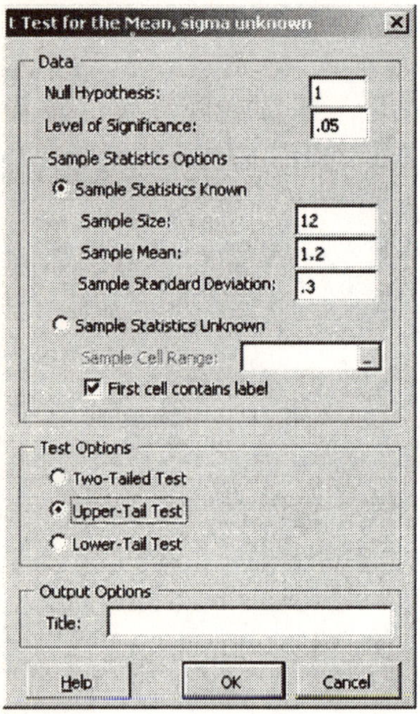

t Test for the Mean, sigma unknown	☒

Data

Null Hypothesis: `1`
Level of Significance: `.05`

Sample Statistics Options
  ◉ Sample Statistics Known
  Sample Size: `12`
  Sample Mean: `1.2`
  Sample Standard Deviation: `.3`
  ○ Sample Statistics Unknown
  Sample Cell Range: [   ] ▫
  ☑ First cell contains label

Test Options
  ○ Two-Tailed Test
  ◉ Upper-Tail Test
  ○ Lower-Tail Test

Output Options
  Title: [   ]

[Help]   [OK]   [Cancel]

Your output should be the same as the output displayed below.

	A	B	C	D	E
1	t Test for Hypothesis of the Mean				
2					
3	Data				
4	Null Hypothesis μ=	1			
5	Level of Significance	0.05			
6	Sample Size	12			
7	Sample Mean	1.2			
8	Sample Standard Deviation	0.3			
9					
10	Intermediate Calculations				
11	Standard Error of the Mean	0.08660254			
12	Degrees of Freedom	11			
13	t Test Statistic	2.309401077			
14					
15	Upper-Tail Test				
16	Upper Critical Value	1.795883691			
17	p-Value	0.020671133			
18	Reject the null hypothesis				
19					
20				Calculations Area	
21				For one-tailed tests:	
22				TDIST value	0.020671
23				1-TDIST value	0.979329

◀

► Exercise 30 (pg. 358)	Testing the Claim That the Mean Number of 8-Oz. Glasses Is Less Than 5.0

If the PHStat add-in has not been loaded, you will need to load it before continuing. Follow the instructions in Section GS 8.2.

---

1. Open the "Ex7_3-30" worksheet in the Chapter 7 folder.

2. At the top of the screen, select **PHStat → One-Sample Tests → t Test for the Mean, sigma unknown**.

3. Complete the t Test for the Mean dialog box as shown below. Click **OK**.

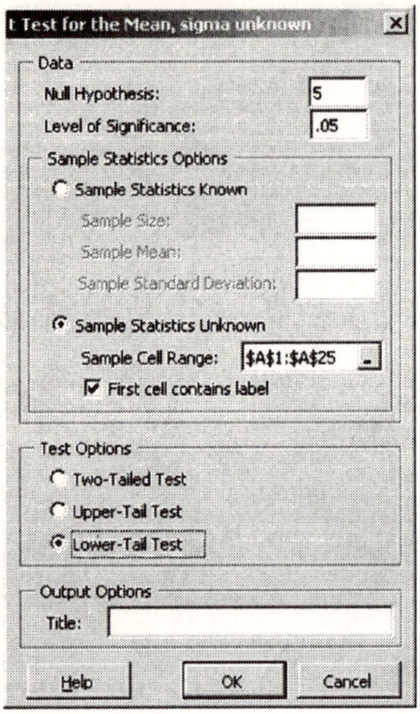

Your output should look like the output displayed below.

	A	B	C	D	E
1	t Test for Hypothesis of the Mean				
2					
3	Data				
4	Null Hypothesis μ=	5			
5	Level of Significance	0.05			
6	Sample Size	24			
7	Sample Mean	4.204166667			
8	Sample Standard Deviation	1.173430188			
9					
10	Intermediate Calculations				
11	Standard Error of the Mean	0.239525434			
12	Degrees of Freedom	23			
13	t Test Statistic	-3.322542077			
14					
15	Lower-Tail Test				
16	Lower Critical Value	-1.713870006			
17	p-Value	0.001482377			
18	Reject the null hypothesis				
19					
20				Calculations Area	
21				For one-tailed tests:	
22				TDIST value	0.001482
23				1-TDIST value	0.998518

Section 7.4

► Example 1 (pg. 341)   Hypothesis Test for a Proportion

If the PHStat add-in has not been loaded, you will need to load it before continuing. Follow the instructions in Section GS 8.2.

1. First open a new Excel worksheet. Then, at the top of the screen, select **PHStat →
    One-Sample Tests → Z Test for the Proportion**.

2. Complete the Z Test for the Proportion dialog box as shown below. Click **OK**.

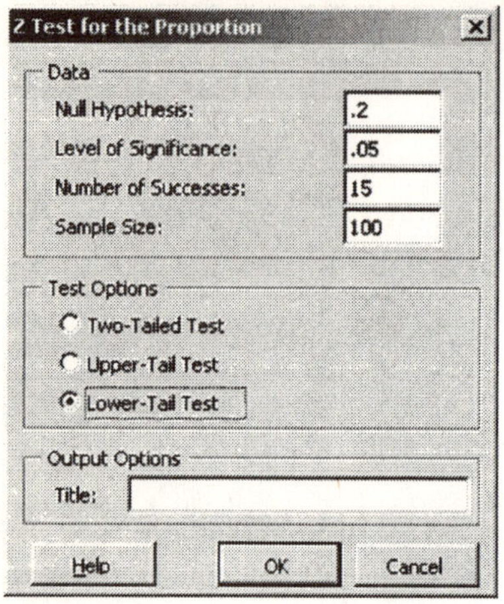

Your output should be the same as the output displayed below.

	A	B	C
1	Z Test of Hypothesis for the Proportion		
2			
3	Data		
4	Null Hypothesis p=	0.2	
5	Level of Significance	0.05	
6	Number of Successes	15	
7	Sample Size	100	
8			
9	Intermediate Calculations		
10	Sample Proportion	0.15	
11	Standard Error	0.04	
12	Z Test Statistic	-1.25	
13			
14	Lower-Tail Test		
15	Lower Critical Value	-1.644853	
16	p-Value	0.105649839	
17	Do not reject the null hypothesis		

◄

► Exercise 9 (pg. 344)   Testing the Claim That More Than 30% of
                          Consumers Have Stopped Buying a Product

If the PHStat add-in has not been loaded, you will need to load it before continuing.
Follow the instructions in Section GS 8.2.

1. First open a new Excel worksheet. Then, at the top of the screen, select **PHStat →
One-Sample Tests → Z Test for the Proportion**.

2. Complete the Z Test for the Proportion dialog box as shown below. Click **OK**.

Note that 32% of 1050 is 336.

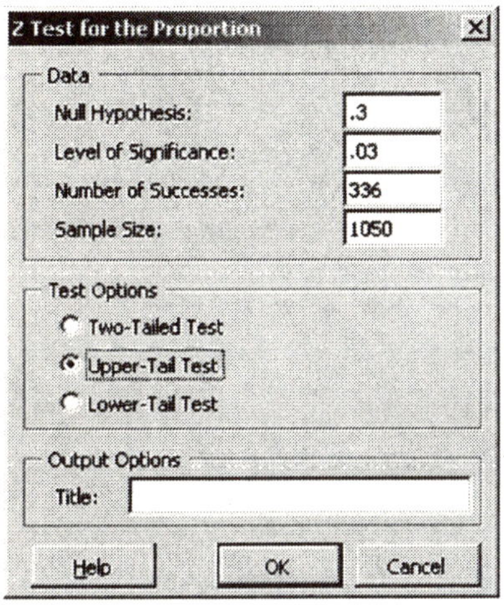

Z Test for the Proportion

Data

Null Hypothesis:	.3
Level of Significance:	.03
Number of Successes:	336
Sample Size:	1050

Test Options

- Two-Tailed Test
- Upper-Tail Test
- Lower-Tail Test

Output Options

Title:

Help | OK | Cancel

Your output should be the same as the output displayed below.

	A	B	C
1	Z Test of Hypothesis for the Proportion		
2			
3	Data		
4	Null Hypothesis $p=$	0.3	
5	Level of Significance	0.03	
6	Number of Successes	336	
7	Sample Size	1050	
8			
9	Intermediate Calculations		
10	Sample Proportion	0.32	
11	Standard Error	0.014142136	
12	Z Test Statistic	1.414213562	
13			
14	Upper-Tail Test		
15	Upper Critical Value	1.880789569	
16	p-Value	0.078649653	
17	Do not reject the null hypothesis		

Technology Lab

▶ Exercise 1 (pg. 366)	Testing the Claim That the Proportion Is Equal to 0.53

If the PHStat add-in has not been loaded, you will need to load it before continuing. Follow the instructions in Section GS 8.2.

1. First open a new Excel worksheet. Then, at the top of the screen, select **PHStat** → **One-Sample Tests** → **Z Test for the Proportion**.

2. Complete the Z Test for the Proportion dialog box as shown below. Click **OK**.

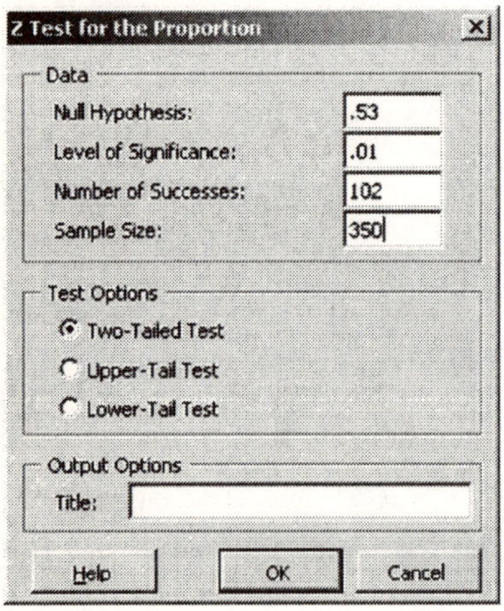

Your output should appear the same as the output shown below.

	A	B	C
1	Z Test of Hypothesis for the Proportion		
2			
3	Data		
4	Null Hypothesis p=	0.53	
5	Level of Significance	0.01	
6	Number of Successes	102	
7	Sample Size	350	
8			
9	Intermediate Calculations		
10	Sample Proportion	0.291428571	
11	Standard Error	0.026677974	
12	Z Test Statistic	-8.942636739	
13			
14	Two-Tailed Test		
15	Lower Critical Value	-2.575834515	
16	Upper Critical value	2.575834515	
17	p-Value	0	
18	Reject the null hypothesis		

◀

Section 7.5

▶ Exercise 27 (pg. 375)

Using the P-Value Method to Perform the Hypothesis Test for Exercise 23

1. Open a new Excel worksheet and click in cell **A1** to place the output there.

2. At the top of the screen, select **Insert → Function**.

3. Under Function category, select **Statistical**. Under Function name, select
    CHIDIST. Click **OK**.

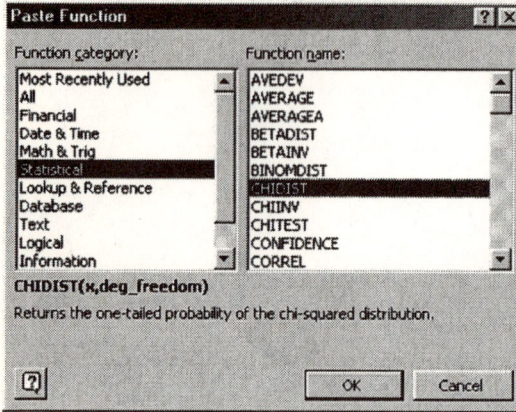

4. Complete the CHIDIST dialog box as shown below. Click **OK**.

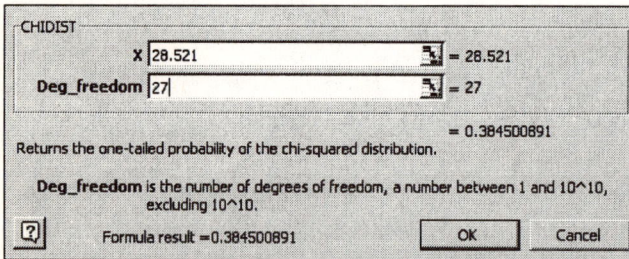

With df=15, the one-tailed probability of X^2 is 0.3845.

	A	B	C	D	E	F	G
1	0.384501						

Hypothesis Testing with Two Samples

Section 8.1

| ► Exercise 19 (pg. 397) | Testing the Claim That Children Ages 3-12 Watched TV More in 1981 Than Today |

1. Open worksheet "Ex8_1-19" in the Chapter 8 folder.

2. Your textbook indicates that the variance of a sample may be used in the z-test formula instead of a known population variance if the sample size is sufficiently large. You will use Excel's Descriptive Statistics tool to obtain the sample variance. At the top of the screen, click **Tools → Data Analysis**. Select **Descriptive Statistics**. Click **OK**.

If Data Analysis does not appear as a choice in the Tools menu, you will need to load the Microsoft Excel ToolPak add-in. Follow the procedure in Section GS 8.1 before continuing.

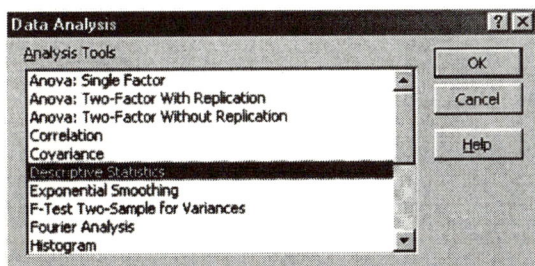

3. Complete the Descriptive Statistics dialog box as shown below. Click **OK**.

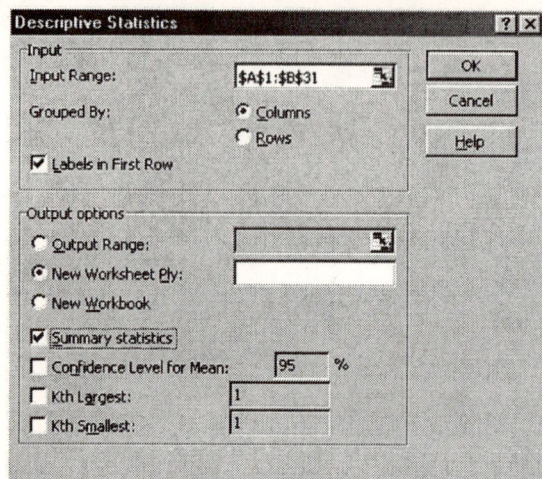

Descriptive Statistics		? ×
Input		
Input Range:	$A$1:$B$31	OK
Grouped By:	⊙ Columns	Cancel
	○ Rows	Help
☑ Labels in First Row		
Output options		
○ Output Range:		
⊙ New Worksheet Ply:		
○ New Workbook		
☑ Summary statistics		
☐ Confidence Level for Mean:	95 %	
☐ Kth Largest:	1	
☐ Kth Smallest:	1	

4. You will want to make column A and column C wider so that the labels are easier to read. The variance of Time A is 0.2401 and the variance of Time B is 0.1075. Return to the worksheet that contains the data. To do this, click on the **Ex8_1-19** tab near the bottom of the screen.

	A	B	C	D
1	*TimeA*		*TimeB*	
2				
3	Mean	2.13	Mean	1.593333
4	Standard Error	0.089462	Standard Error	0.059872
5	Median	2.1	Median	1.6
6	Mode	2.1	Mode	1.6
7	Standard Deviation	0.490004	Standard Deviation	0.327933
8	Sample Variance	0.240103	Sample Variance	0.10754
9	Kurtosis	0.863836	Kurtosis	-0.57822
10	Skewness	-0.00957	Skewness	-0.25732
11	Range	2.3	Range	1.3
12	Minimum	1	Minimum	0.9
13	Maximum	3.3	Maximum	2.2
14	Sum	63.9	Sum	47.8
15	Count	30	Count	30

5. At the top of the screen, click **Tools → Data Analysis**. Select **z-Test: Two Sample for Means** and click **OK**.

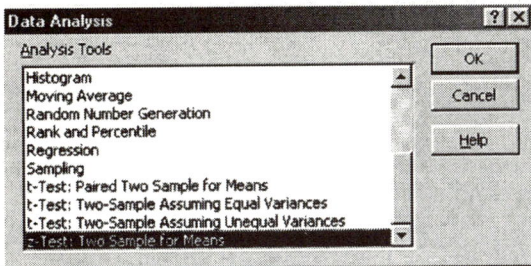

6. Complete the z-Test: Two Sample for Means dialog box as shown below. Click **OK**.

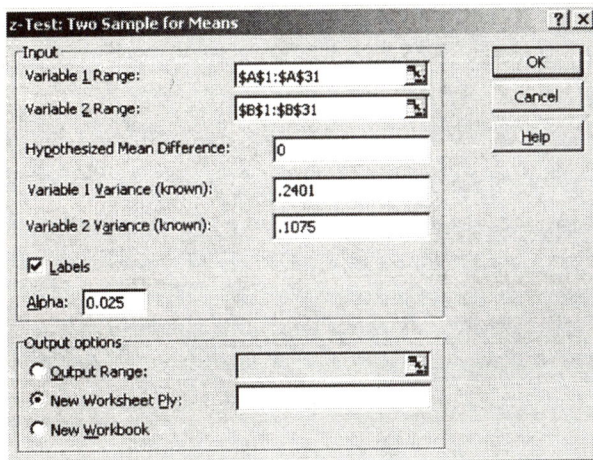

The output is displayed in a new worksheet. You will want to adjust the width of column A so that you can read the labels. Your output should look similar to the output displayed below.

	A	B	C
1	z-Test: Two Sample for Means		
2			
3		TimeA	TimeB
4	Mean	2.13	1.593333
5	Known Variance	0.2401	0.1075
6	Observations	30	30
7	Hypothesized Mean Difference	0	
8	z	4.985691	
9	P(Z<=z) one-tail	3.09E-07	
10	z Critical one-tail	1.959961	
11	P(Z<=z) two-tail	6.18E-07	
12	z Critical two-tail	2.241395	

◀

Section 8.2

▶ Exercise 15 (pg. 407)   Testing the Claim That the Mean Footwell Intrusion for Small & Midsize Cars Is Equal

If the PHStat add-in has not been loaded, you will need to load it before continuing. Follow the instructions in Section GS 8.2.

1. First open a new Excel worksheet. Then, at the top of the screen, select **PHStat →Two-Sample Tests → t Test for Differences in Two Means**.

2. Complete the t Test for Differences in Two Means dialog box as shown below. Click **OK**.

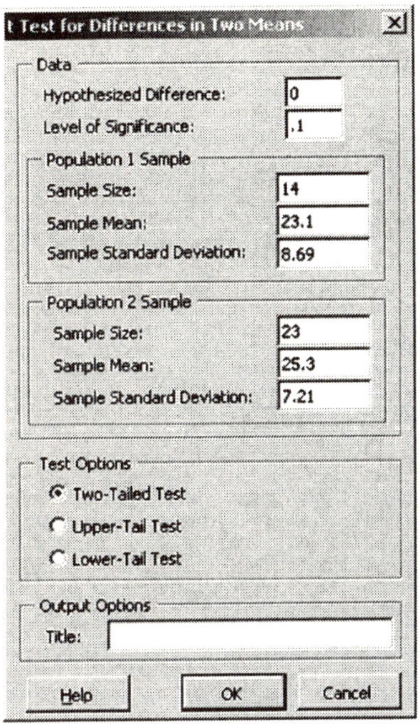

t Test for Differences in Two Means	☒

Data

Hypothesized Difference: `0`

Level of Significance: `.1`

Population 1 Sample

Sample Size: `14`

Sample Mean: `23.1`

Sample Standard Deviation: `8.69`

Population 2 Sample

Sample Size: `23`

Sample Mean: `25.3`

Sample Standard Deviation: `7.21`

Test Options

⦿ Two-Tailed Test

○ Upper-Tail Test

○ Lower-Tail Test

Output Options

Title: `_____`

Help	OK	Cancel

The output is displayed in a new worksheet named "Hypothesis." You may need to scroll down a couple rows to see all of it.

	A	B
1	**t Test for Differences in Two Means**	
2		
3	**Data**	
4	**Hypothesized Difference**	0
5	**Level of Significance**	0.1
6	**Population 1 Sample**	
7	**Sample Size**	14
8	**Sample Mean**	23.1
9	**Sample Standard Deviation**	8.69
10	**Population 2 Sample**	
11	**Sample Size**	23
12	**Sample Mean**	25.3
13	**Sample Standard Deviation**	7.21
14		
15	Intermediate Calculations	
16	Population 1 Sample Degrees of Freedom	13
17	Population 2 Sample Degrees of Freedom	22
18	Total Degrees of Freedom	35
19	Pooled Variance	60.72456
20	Difference in Sample Means	-2.2
21	*t*-Test Statistic	-0.83285
22		
23	**Two-Tailed Test**	
24	**Lower Critical Value**	-1.68957
25	**Upper Critical Value**	1.689573

‖◄ ◄ ► ►‖ \ **Hypothesis** / Sheet1 / Sheet2 / Sheet3 /

Section 8.3

▶ Example 2 (pg. 414)	The t-Test for the Difference Between Means

1. Open a new Excel worksheet and enter the golfers' scores as shown below.

	A	B
1	Old Design	New Design
2	89	83
3	84	83
4	96	92
5	82	84
6	74	76
7	92	91
8	85	80
9	91	91

2. At the top of the screen, click **Tools → Data Analysis**. Select **t-Test: Paired Two Sample for Means**. Click **OK**.

If Data Analysis does not appear as a choice in the Tools menu, you will need to load the Microsoft Excel ToolPak add-in. Follow the procedure in Section GS 8.1 before continuing.

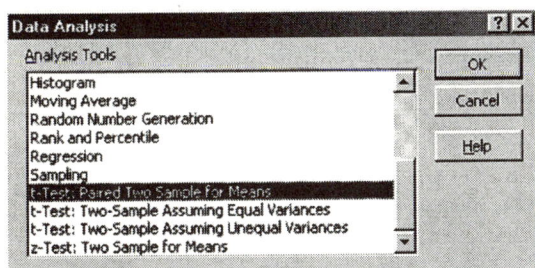

Data Analysis

Analysis Tools

Histogram
Moving Average
Random Number Generation
Rank and Percentile
Regression
Sampling
t-Test: Paired Two Sample for Means
t-Test: Two-Sample Assuming Equal Variances
t-Test: Two-Sample Assuming Unequal Variances
z-Test: Two Sample for Means

OK
Cancel
Help

3.   Complete the t-Test: Paired Two Sample for Means dialog box as shown below. Click **OK**.

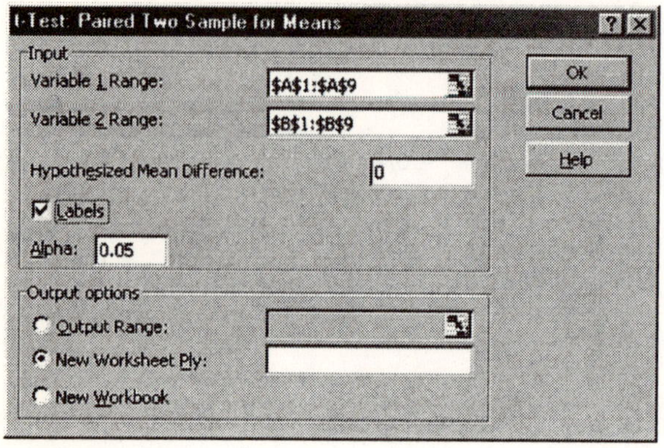

The output is displayed in a new worksheet. You will want to adjust the width of the columns so that you can read all the labels. Your output should look similar to the output displayed below.

	A	B	C
1	t-Test: Paired Two Sample for Means		
2			
3		Old Design	New Design
4	Mean	86.625	85
5	Variance	47.410714	33.714286
6	Observations	8	8
7	Pearson Correlation	0.896872	
8	Hypothesized Mean Difference	0	
9	df	7	
10	t Stat	1.4982596	
11	P(T<=t) one-tail	0.0888692	
12	t Critical one-tail	1.8945775	
13	P(T<=t) two-tail	0.1777385	
14	t Critical two-tail	2.3646226	

◀

| ▶ Exercise 17 (pg. 418) | Testing the Hypothesis That Verbal SAT Scores Improved the Second Time |

1.   Open worksheet "Ex8_3-17" in the Chapter 8 folder.

2. At the top of the screen, click **Tools → Data Analysis**. Select **t-Test: Paired Two Sample for Means**. Click **OK**.

If Data Analysis does not appear as a choice in the Tools menu, you will need to load the Microsoft Excel ToolPak add-in. Follow the procedure in Section GS 8.1 before continuing.

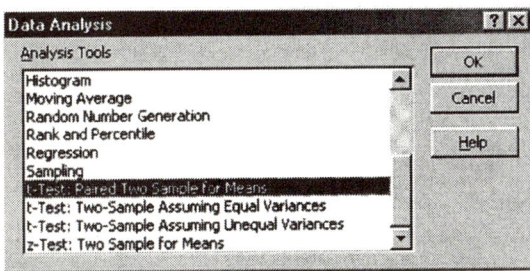

3. Complete the t-Test: Paired Two Sample for Means dialog box as shown below. Click **OK**.

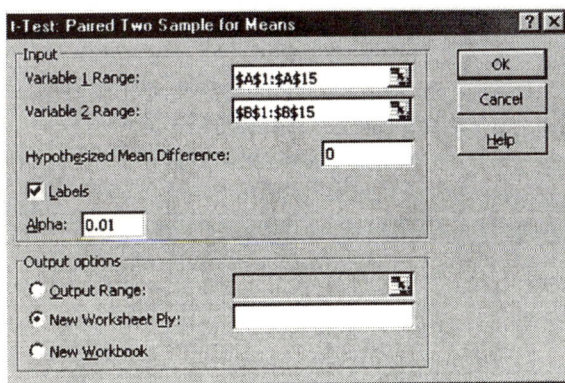

The output is displayed in a new worksheet. You will want to make the columns wider so that you can read all the labels. Your output should appear similar to the output shown below.

	A	B	C
1	t-Test: Paired Two Sample for Means		
2			
3		First SAT	Second SAT
4	Mean	483.6429	517.357143
5	Variance	8372.555	5060.09341
6	Observations	14	14
7	Pearson Correlation	0.896142	
8	Hypothesized Mean Difference	0	
9	df	13	
10	t Stat	-3.0011	
11	P(T<=t) one-tail	0.005109	
12	t Critical one-tail	2.650304	
13	P(T<=t) two-tail	0.010217	
14	t Critical two-tail	3.012283	

◀

▶ Exercise 25 (pg. 421)   Constructing a 90% Confidence Interval for μ_D, the Mean Increase in Hours of Sleep

1. Open a new Excel worksheet and enter the data as shown below.

	A	B	C
1	Without drug	With new drug	
2	1.8	3	
3	2	3.6	
4	3.4	4	
5	3.5	4.4	
6	3.7	4.5	
7	3.8	5.2	
8	3.9	5.5	
9	3.9	5.7	
10	4	6.2	
11	4.9	6.3	
12	5.1	6.6	
13	5.2	7.8	
14	5	7.2	
15	4.5	6.5	
16	4.2	5.6	
17	4.7	5.9	

2. Use Excel to calculate a difference score for each of the 16 patients. First, click in cell D1 and key in the label **Difference**.

	A	B	C	D
1	Without drug	With new drug		Difference
2	1.8	3		

3. Click in cell D2 and enter the formula **=A2-B2** as shown below. Press [**Enter**].

	A	B	C	D	E
1	Without drug	With new drug		Difference	
2	1.8	3		=A2-B2	

4. Click in cell **D2** (where –1.2 now appears) and copy the contents of that cell to cells D3 through D17.

	A	B	C	D
1	Without drug	With new drug		Difference
2	1.8	3		-1.2
3	2	3.6		-1.6
4	3.4	4		-0.6
5	3.5	4.4		-0.9
6	3.7	4.5		-0.8
7	3.8	5.2		-1.4
8	3.9	5.5		-1.6
9	3.9	5.7		-1.8
10	4	6.2		-2.2
11	4.9	6.3		-1.4
12	5.1	6.6		-1.5
13	5.2	7.8		-2.6
14	5	7.2		-2.2
15	4.5	6.5		-2
16	4.2	5.6		-1.4
17	4.7	5.9		-1.2

5. You will now obtain descriptive statistics for Difference. At the top of the screen, click **Tools → Data Analysis**. Select **Descriptive Statistics**. Click **OK**.

If Data Analysis does not appear as a choice in the Tools menu, you will need to load the Microsoft Excel ToolPak add-in. Follow the procedure in Section GS 8.1 before continuing.

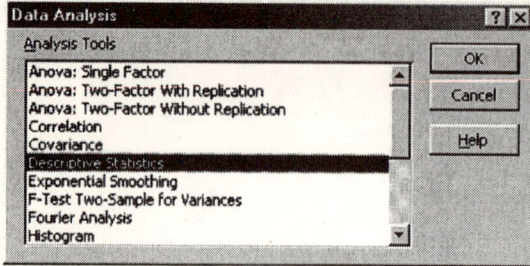

6.   Complete the Descriptive Statistics dialog box as shown below. Click **OK**.

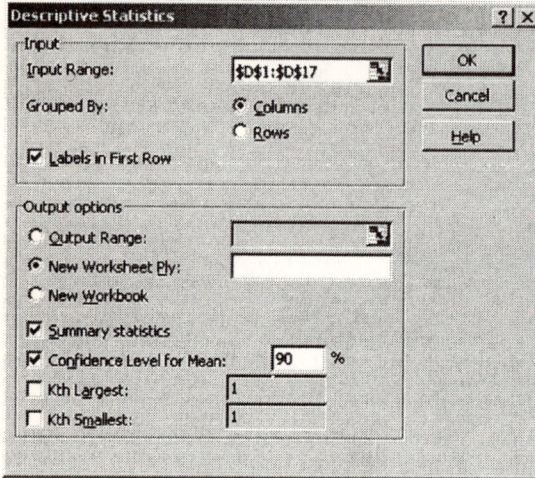

The output is displayed in a new worksheet. To calculate the lower limit of the 90% confidence interval, subtract 0.2376 from –1.525. To calculate the upper limit, add 0.2376 to –1.525.

	A	B
1	*Difference*	
2		
3	Mean	-1.525
4	Standard Error	0.135554
5	Median	-1.45
6	Mode	-1.6
7	Standard Deviation	0.542218
8	Sample Variance	0.294
9	Kurtosis	-0.26069
10	Skewness	-0.23658
11	Range	2
12	Minimum	-2.6
13	Maximum	-0.6
14	Sum	-24.4
15	Count	16
16	Confidence Level(90.0%)	0.237634

◀

Section 8.4

▶ Example 1 (pg. 424)	A Two-Sample z-Test for the Difference Between Proportions

If the PHStat add-in has not been loaded, you will need to load it before continuing. Follow the instructions in Section GS 8.2.

1. First open a new Excel worksheet. Then, at the top of the screen, select **PHStat → Two-Sample Tests → Z Test for Differences in Two Proportions**.

2. Complete the Z Test for Differences in Two Proportions dialog box as shown below. Click **OK**.

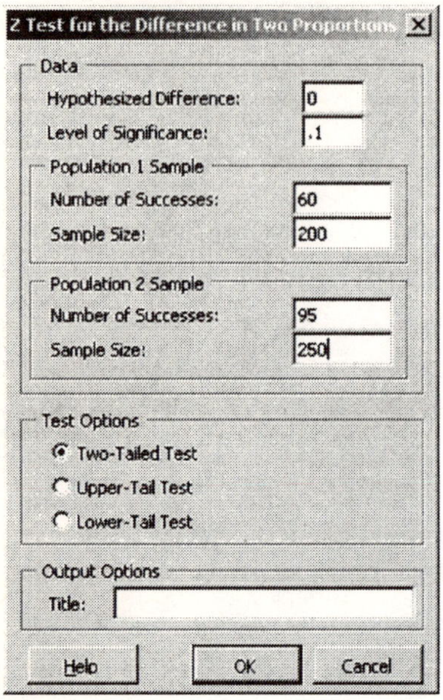

The output is displayed in a worksheet named "Hypothesis."

	A	B
1	**Z Test for Differences in Two Proportions**	
2		
3	**Data**	
4	**Hypothesized Difference**	0
5	**Level of Significance**	0.1
6	**Group 1**	
7	**Number of Successes**	60
8	**Sample Size**	200
9	**Group 2**	
10	**Number of Successes**	95
11	**Sample Size**	250
12		
13	Intermediate Calculations	
14	Group 1 Proportion	0.3
15	Group 2 Proportion	0.38
16	Difference in Two Proportions	-0.08
17	Average Proportion	0.344444444
18	Z Test Statistic	-1.774615984
19		
20	**Two-Tailed Test**	
21	**Lower Critical Value**	-1.644853
22	**Upper Critical Value**	1.644853
23	*p*-Value	0.075961219
24	**Reject the null hypothesis**	

◀

► **Exercise 9   (pg. 426)** | **Testing the Claim That the Use of Alternative Medicines Has Not Changed**

If the PHStat add-in has not been loaded, you will need to load it before continuing. Follow the instructions in Section GS 8.2.

---

1. First open a new Excel worksheet. Then, at the top of the screen, select **PHStat** → **Two-Sample Tests** → **Z Test for Differences in Two Proportions**.

2. Complete the Z Test for Differences in Two Proportions dialog box as shown below. Click **OK**.

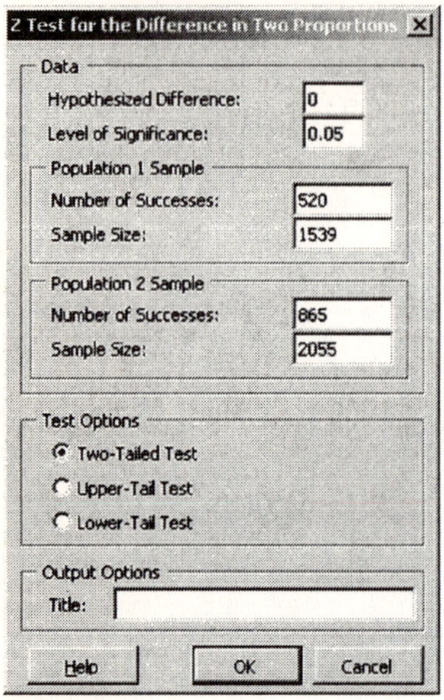

Z Test for the Difference in Two Proportions	
Data	
Hypothesized Difference:	0
Level of Significance:	0.05
Population 1 Sample	
Number of Successes:	520
Sample Size:	1539
Population 2 Sample	
Number of Successes:	865
Sample Size:	2055
Test Options	
⦿ Two-Tailed Test	
○ Upper-Tail Test	
○ Lower-Tail Test	
Output Options	
Title:	
Help	OK	Cancel

The output is displayed in a worksheet named "Hypothesis."

	A	B
1	Z Test for Differences in Two Proportions	
2		
3	Data	
4	Hypothesized Difference	0
5	Level of Significance	0.05
6	Group 1	
7	Number of Successes	520
8	Sample Size	1539
9	Group 2	
10	Number of Successes	865
11	Sample Size	2055
12		
13	Intermediate Calculations	
14	Group 1 Proportion	0.337881741
15	Group 2 Proportion	0.420924574
16	Difference in Two Proportions	-0.083042833
17	Average Proportion	0.385364496
18	Z Test Statistic	-5.06166817
19		
20	Two-Tailed Test	
21	Lower Critical Value	-1.959961082
22	Upper Critical Value	1.959961082
23	p -Value	4.16302E-07
24	Reject the null hypothesis	

Technology Lab

► Exercise 1 (pg. 429)	Testing the Hypothesis That the Probability of a "Found Coin" Lying Heads Up is 0.5

If the PHStat add-in has not been loaded, you will need to load it before continuing. Follow the instructions in Section GS 8.2.

1. First open a new Excel worksheet. Then, at the top of the screen, select **PHStat →
   One-Sample Tests → Z Test for the Proportion**.

2. Complete the Z Test for the Proportion dialog box as shown below. Click **OK**.

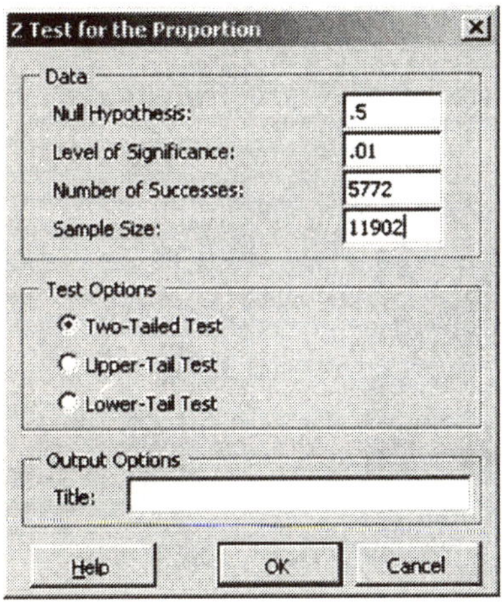

Your output should look similar to the output displayed below.

	A	B	C
1	Z Test of Hypothesis for the Proportion		
2			
3	Data		
4	Null Hypothesis *p*=	0.5	
5	Level of Significance	0.01	
6	Number of Successes	5772	
7	Sample Size	11902	
8			
9	Intermediate Calculations		
10	Sample Proportion	0.484960511	
11	Standard Error	0.004583107	
12	Z Test Statistic	-3.281504874	
13			
14	Two-Tailed Test		
15	Lower Critical Value	-2.575834515	
16	Upper Critical value	2.575834515	
17	*p*-Value	0.001032667	
18	Reject the null hypothesis		

▶ Exercise 3 (pg. 429)   Simulating "Tails Over Heads"
_____

1. Open a new Excel worksheet and click in cell **A1** to place the output there.

2. At the top of the screen, select **Tools → Data Analysis**. Select **Random Number Generation**. Click **OK**.

3.   Complete the Random Number Generation dialog box as shown below. Click **OK**.

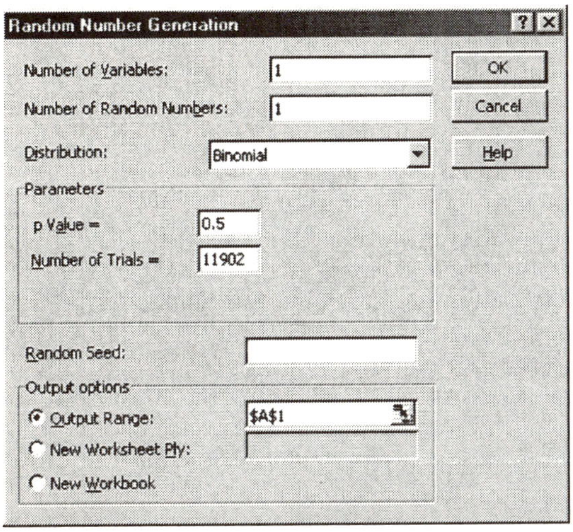

Random Number Generation		? X
Number of Variables:	1	OK
Number of Random Numbers:	1	Cancel
Distribution:	Binomial ▼	Help

Parameters

p Value =	0.5
Number of Trials =	11902

Random Seed: [   ]

Output options

(•) Output Range: $A$1
( ) New Worksheet Ply:
( ) New Workbook

The output for this simulation indicates that 5,955 of the 11,902 coins were found heads up. Because this output was generated randomly, it is unlikely that your output will be exactly the same.

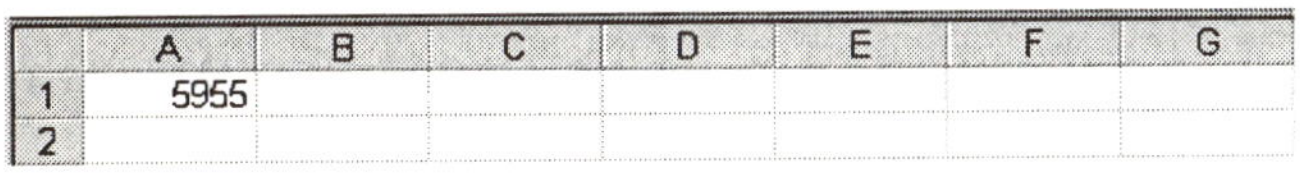

	A	B	C	D	E	F	G
1	5955						
2							

Correlation and
Regression

Section 9.1

▶ Example 3 (pg. 444) | Constructing a Scatter Plot

1. Open worksheet "OldFaithful" in the Chapter 9 folder.

2. Click on any cell within the table of data. At the top of the screen, click **Insert** and select **Chart** from the menu that appears.

3. In the Chart Type dialog box, select the **XY (Scatter)** chart type by clicking on it. Then click on the topmost chart sub-type. Click **Next**.

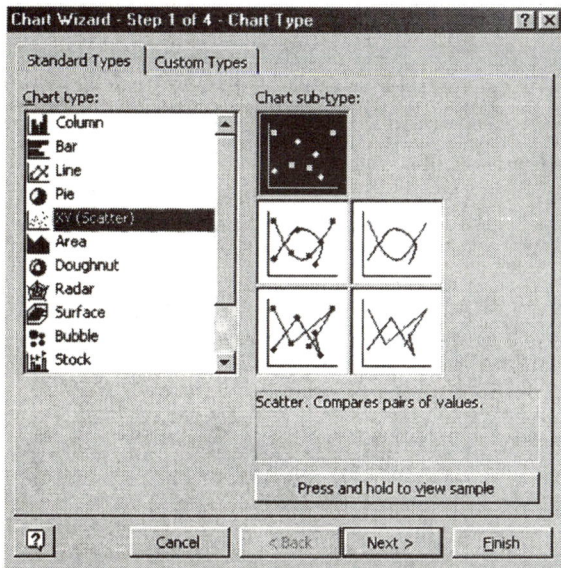

4. In the Chart Source Data dialog box, the entry in the data range field should be **=OldFaithful!$A$1:$B$36**. Make any necessary corrections. Then click **Next**.

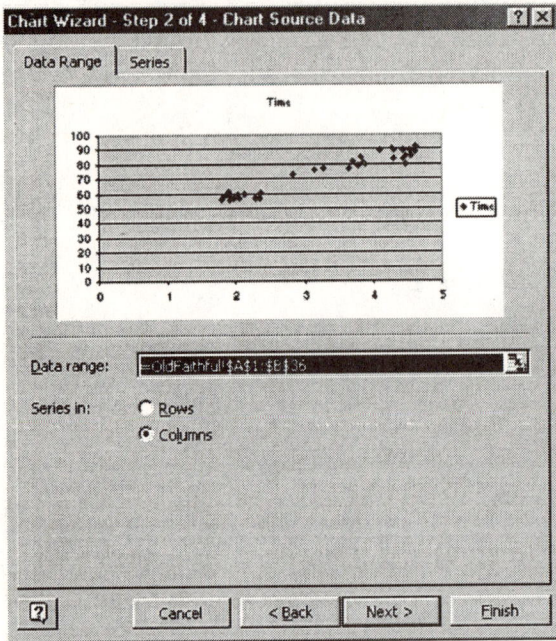

5. Click the **Titles** tab at the top of the Chart Options dialog box. In the Chart title field, change the title to **Duration of Old Faithful's Eruptions by Time Until the Next Eruption**. In the Value (X) axis field, enter **Duration (in minutes)**. In the Value (Y) axis field, enter **Time until the next eruption (in minutes)**.

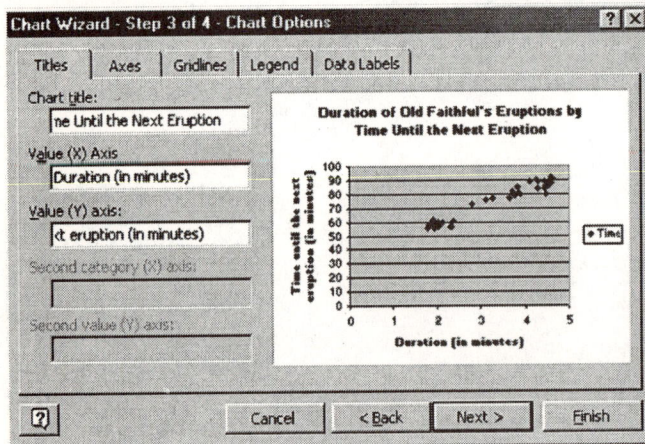

6. Click the **Legend** tab at the top of the Chart Options dialog box. Click in the box to the left of **Show Legend** so that a checkmark does not appear there. This removes the Time legend from the right side of the scatter plot. Click **Next**.

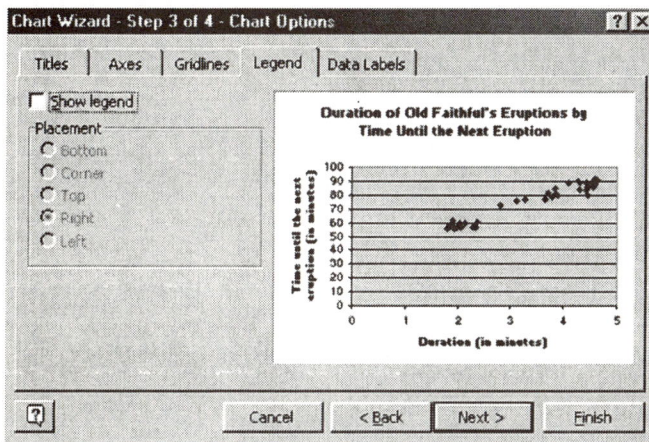

7. The Chart Location dialog box presents two options for placement of the scatter plot. For this example, select **As object in** the same worksheet as the data. To do this, click the button to the left of **As object in** so that a black dot appears there. Click **Finish**.

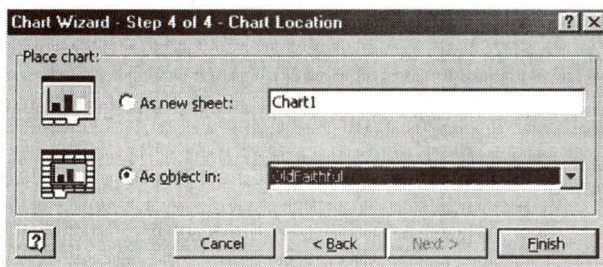

8. Make the chart taller so that it is easier to read. To do this, first click within the figure near a border. Black square handles appear. Then click on the center handle on the bottom border of the figure and drag it down a few rows.

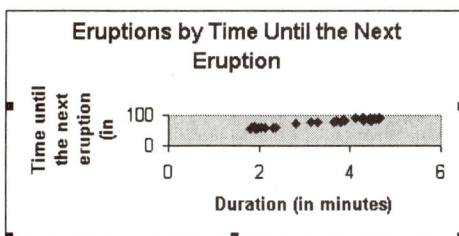

Your scatter plot should now look similar to the one shown below.

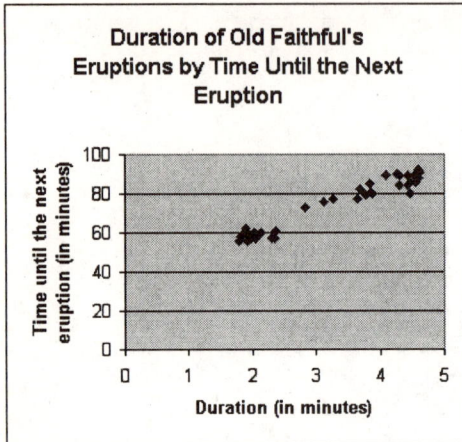

Duration of Old Faithful's Eruptions by Time Until the Next Eruption

▶ Example 5 (pg. 447)	Using Technology to Find a Correlation Coefficient

1. Open worksheet "OldFaithful" in the Chapter 9 folder and click in cell C37 to place the output there.

	A	B	C	D	E	F	G
35	4.6	92					
36	4.63	91					
37							

2. At the top of the screen, click **Insert → Function**.

3. Under Function category, select **Statistical**. Under Function name, select **CORREL**. Click **OK**.

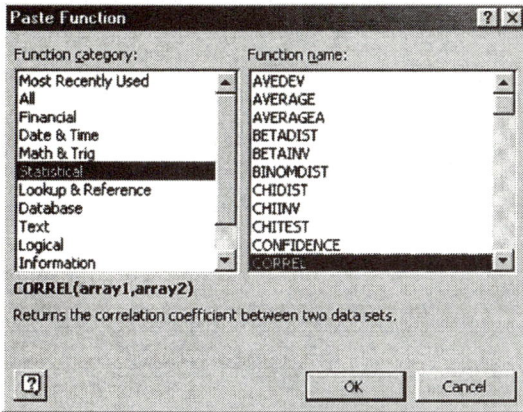

4. Complete the CORREL dialog box as shown below. Click **OK**.

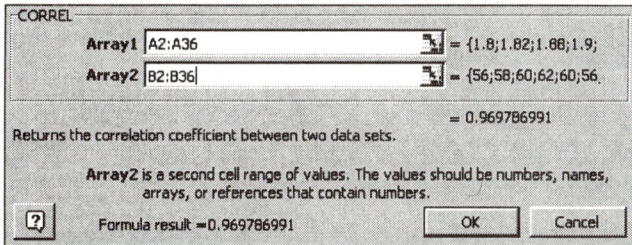

Your output should look like the output shown below.

	A	B	C	D	E	F	G
35	4.6	92					
36	4.63	91					
37			0.969787				

► Exercise 13 (pg. 454)	Constructing a Scatter Plot and Determining Correlation

1. Open worksheet "Ex9_1-13" in the Chapter 9 folder.

2. Click on any cell within the table of data. At the top of the screen, click **Insert** and select **Chart** from the menu that appears.

3. In the Chart Type dialog box, select the **XY (Scatter)** chart type by clicking on it. Then click on the topmost chart sub-type. Click **Next**.

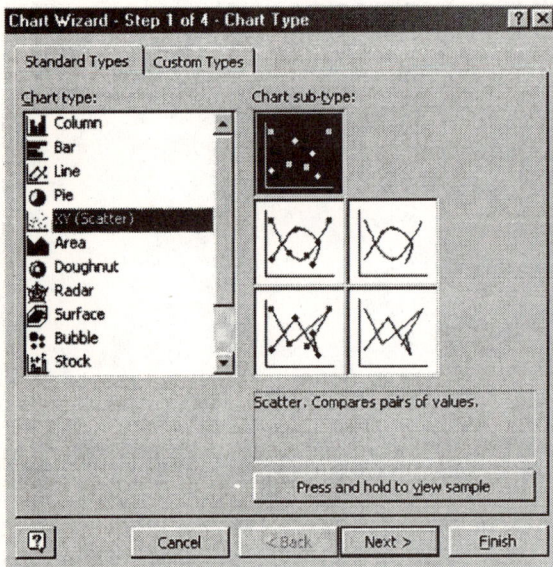

4. In the Chart Source Data dialog box, the entry in the data range field should be ='Ex9_1-13'!$A$1:$B$14. Make any necessary corrections. Then click **Next**.

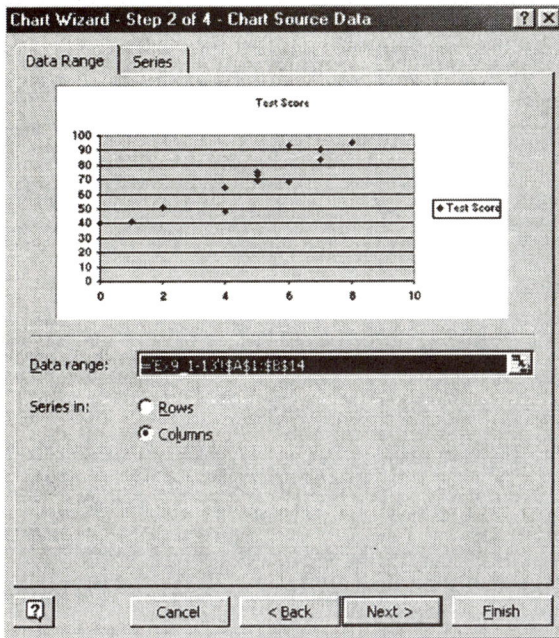

Chart Wizard - Step 2 of 4 - Chart Source Data

Data Range | Series

Test Score

Data range: ='Ex9_1-13'!$A$1:$B$14

Series in: ○ Rows ● Columns

Cancel | < Back | Next > | Finish

5. Click the **Titles** tab at the top of the Chart Options dialog box. In the Chart title field, change the title to **Hours Spent Studying for a Test by Test Score**. In the Value (X) axis field, enter **Hours**. In the Value (Y) axis field, enter **Test Score**.

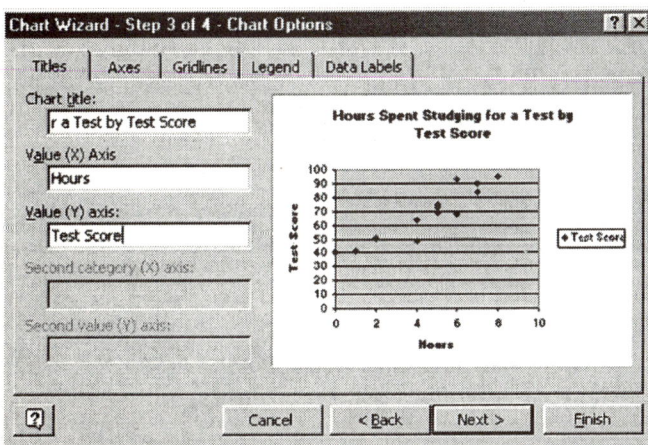

Chart Wizard - Step 3 of 4 - Chart Options

Titles | Axes | Gridlines | Legend | Data Labels

Chart title: r a Test by Test Score

Value (X) Axis: Hours

Value (Y) axis: Test Score

Second category (X) axis:

Second value (Y) axis:

Hours Spent Studying for a Test by Test Score

Cancel | < Back | Next > | Finish

6. Click the **Legend** tab at the top of the Chart Options dialog box. Click in the box to the left of **Show Legend** so that a checkmark does not appear there. This removes the Test Score legend from the right side of the scatter plot. Click **Next**.

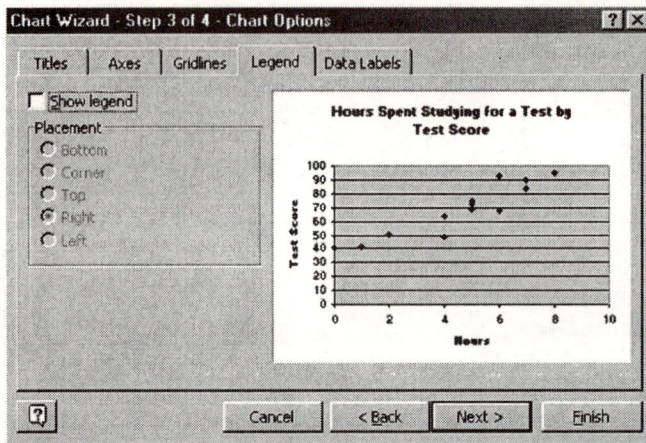

7. The Chart Location dialog box presents two options for placement of the scatter plot. For this example, select **As object in** the same worksheet as the data. To do this, click the button to the left of **As object in** so that a black dot appears there. Click **Finish**.

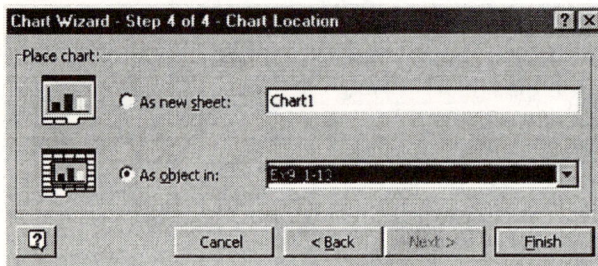

8. Make the chart taller so that it is easier to read. To do this, first click within the figure near a border. Black square handles appear. Then click on the center handle on the bottom border of the figure and drag it down a few rows.

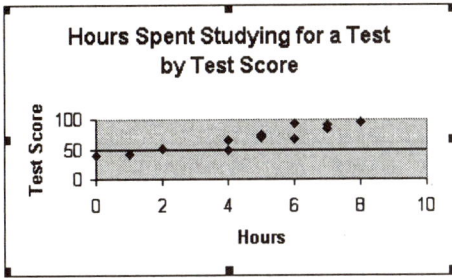

Your scatter plot should now look similar to the one shown below.

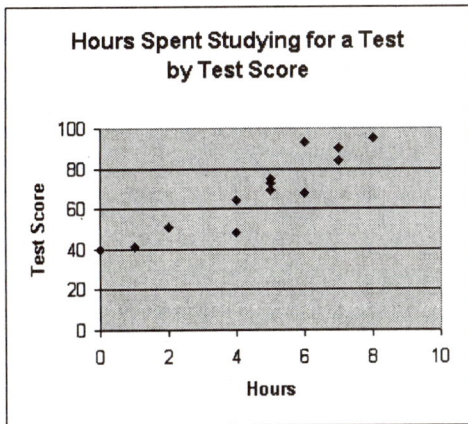

9. You will now use Excel to find the correlation between hours and test score. Click in cell **B16** of the worksheet to place the output there.

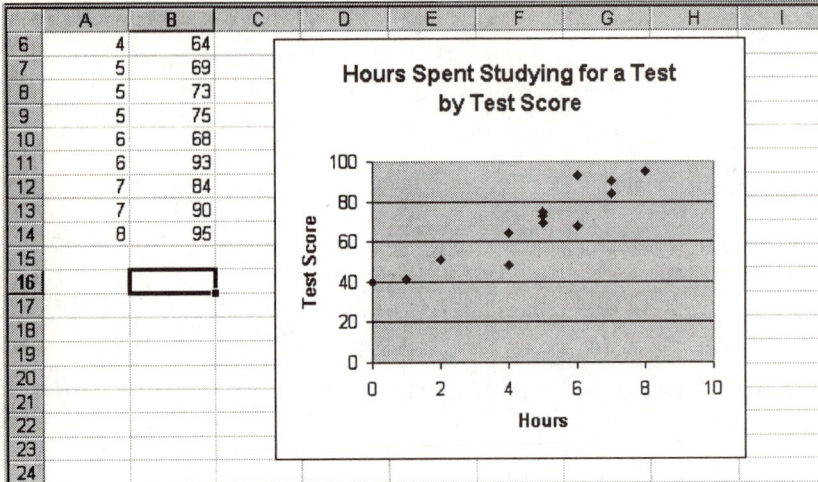

10. At the top of the screen, click **Insert** → **Function**.

11. Under Function category, select **Statistical**. Under Function name, select **CORREL**. Click **OK**.

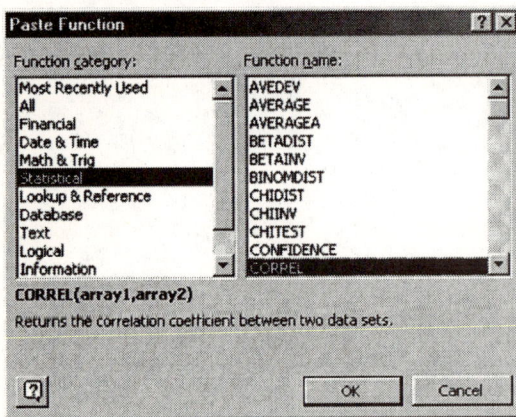

12. Complete the CORREL dialog box as shown below. Click **OK**.

Your output should look like the output shown below.

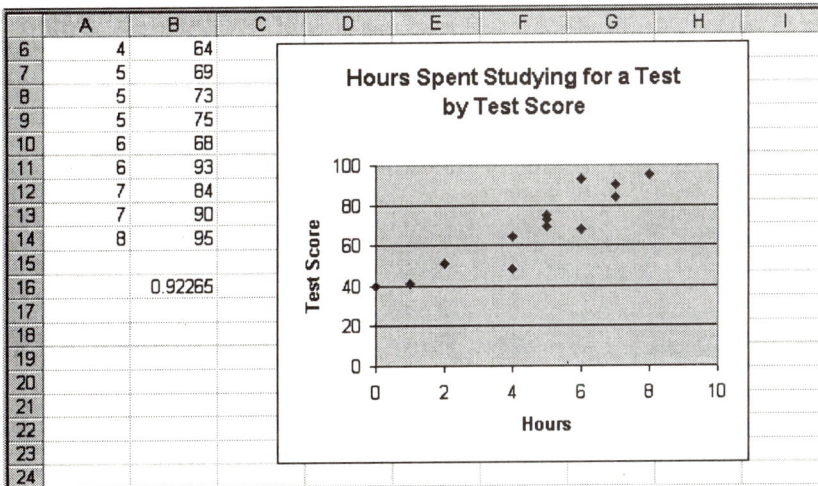

Section 9.2

▶ Example 2 (pg. 460)	Using Technology to Find a Regression Equation

1. Open worksheet "OldFaithful" in the Chapter 9 folder.

2. At the top of the screen, click **Tools** → **Data Analysis**. Select **Regression** and click **OK**.

If Data Analysis does not appear as a choice in the Tools menu, you will need to load the Microsoft Excel Analysis ToolPak add-in. Follow the procedure in Section GS 8.1 before continuing.

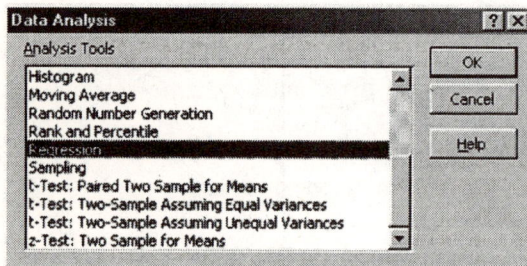

3. Complete the Regression dialog box as shown below. Click **OK**.

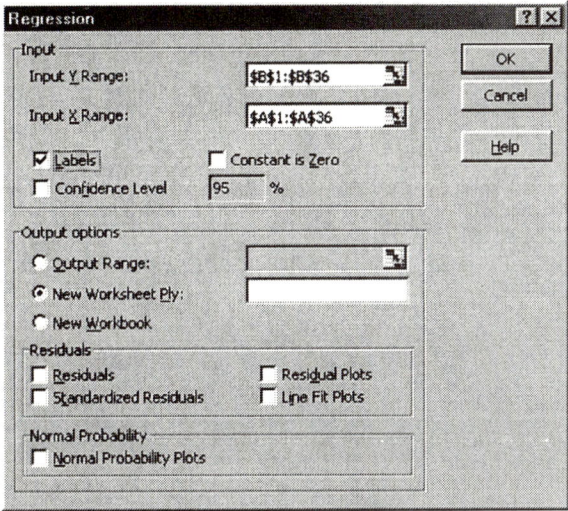

You will want to make some of the columns wider so that you can read all the labels. Your output should look similar to the output shown below. The intercept and slope of the regression equation are shown in the bottom two lines of the output under the label "Coefficients." The intercept is 35.301 and the slope is 11.824.

	A	B	C	D	E	F	G	H	I
16		Coefficients	Standard Error	t Stat	P-value	Lower 95%	Upper 95%	Lower 95.0%	Upper 95.0%
17	Intercept	35.301171	1.804267929	19.56537	1E-19	31.630357	38.97198	31.630357	38.9719847
18	Duration	11.824408	0.517788852	22.83635	8.57E-22	10.770958	12.87786	10.770958	12.877858

▶ Exercise 11 (pg. 463)	Finding the Equation of a Regression Line

1. Open a new Excel worksheet and enter the age and systolic blood pressure data as shown below.

	A	B	C	D	E	F	G
1	Age	Blood Pressure					
2	16	109					
3	25	122					
4	39	143					
5	45	132					
6	49	199					
7	64	185					
8	70	199					

2. At the top of the screen, click **Tools** → **Data Analysis**. Select **Regression** and click **OK**.

If Data Analysis does not appear as a choice in the Tools menu, you will need to load the Microsoft Excel Analysis ToolPak add-in. Follow the procedure in Section GS 8.1 before continuing.

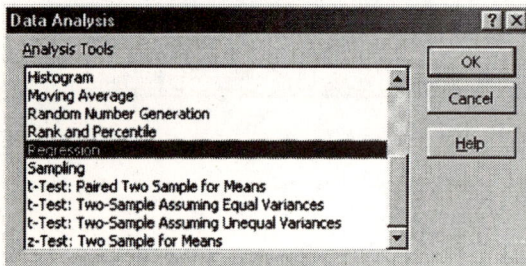

Data Analysis

Analysis Tools

Histogram
Moving Average
Random Number Generation
Rank and Percentile
Regression
Sampling
t-Test: Paired Two Sample for Means
t-Test: Two-Sample Assuming Equal Variances
t-Test: Two-Sample Assuming Unequal Variances
z-Test: Two Sample for Means

OK
Cancel
Help

3. Complete the Regression dialog box as shown below. Click **OK**.

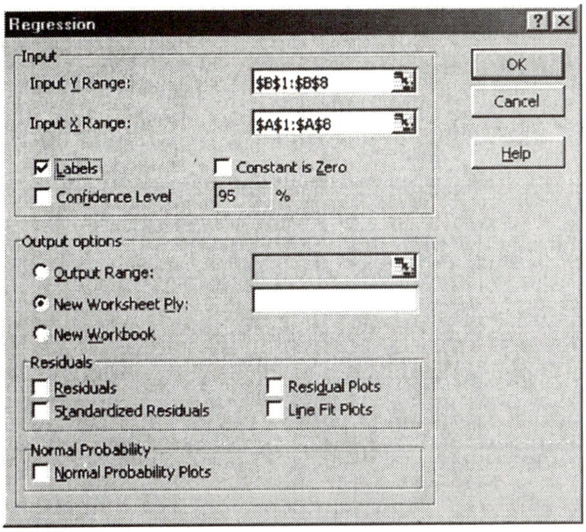

You will want to make some of the columns wider so that you can read all the labels. Your output should look similar to the output shown below. The intercept and slope of the regression equation are shown in the bottom two lines of the output under the label "Coefficients." The intercept is 79.733 and the slope is 1.724.

	A	B	C	D	E	F	G	H	I
16		Coefficients	Standard Error	t Stat	P-value	Lower 95%	Upper 95%	Lower 95.0%	Upper 95.0%
17	Intercept	79.7334	19.43492651	4.102583	0.009331	29.77441	129.6924	29.774413	129.692388
18	Age	1.7235915	0.408765078	4.216582	0.008355	0.672829	2.774354	0.67282918	2.77435392

Section 9.3

▶ Example 2 (pg. 471)	Finding the Standard Error of Estimate

1. Open a new Excel worksheet and enter the expenditures and sales data as shown below.

	A	B	C	D	E	F	G
1	Expenditures	Sales					
2	2.4	225					
3	1.6	184					
4	2	220					
5	2.6	240					
6	1.4	180					
7	1.6	184					
8	2	186					
9	2.2	215					

2. At the top of the screen, click **Tools → Data Analysis**. Select **Regression** and click **OK**.

If Data Analysis does not appear as a choice in the Tools menu, you will need to load the Microsoft Excel Analysis ToolPak add-in. Follow the procedure in Section GS 8.1 before continuing.

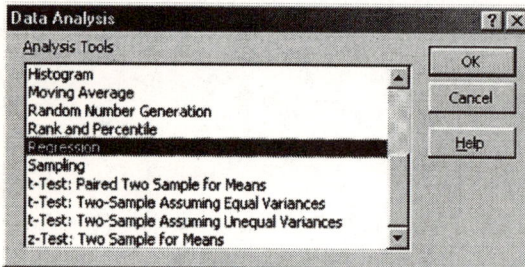

3. Complete the Regression dialog box as shown below. Click **OK**.

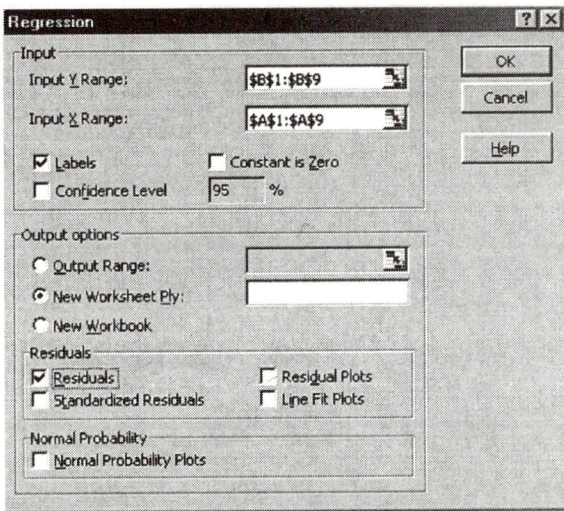

The standard error of estimate is displayed in the upper portion of the output. You will want to widen column A so that you can read all the labels. The standard error of estimate is equal to 10.2903.

	A	B
1	SUMMARY OUTPUT	
2		
3	*Regression Statistics*	
4	Multiple R	0.91290528
5	R Square	0.83339606
6	Adjusted R Square	0.80562874
7	Standard Error	10.2903201
8	Observations	8

Section 9.4

| ▶ Example 1 (pg. 478) | Finding a Multiple Regression Equation |

1. Open worksheet "Salary" in the Chapter 9 folder.

2. At the top of the screen, click **Tools** → **Data Analysis**. Select **Regression** and click **OK**.

If Data Analysis does not appear as a choice in the Tools menu, you will need to load the Microsoft Excel Analysis ToolPak add-in. Follow the procedure in Section GS 8.1 before continuing.

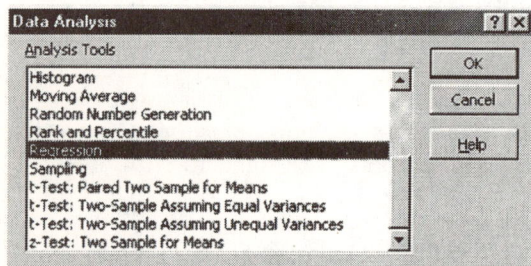

3. Complete the Regression dialog box as shown below. Click **OK**.

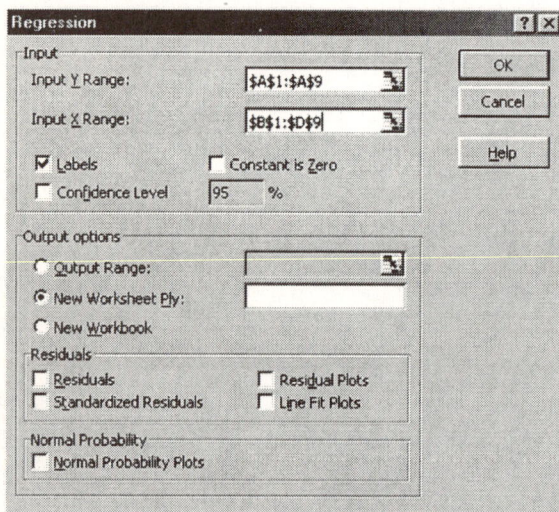

The coefficients for the multiple regression equation are displayed in the lower portion of the output under the label "Coefficients."

	A	B	C	D	E	F	G	H	I
16		Coefficients	Standard Error	t Stat	P-value	Lower 95%	Upper 95%	Lower 95.0%	Upper 95.0%
17	Intercept	29764.446	1981.346465	15.02233	0.000114	24263.33	35265.56	24263.3349	35265.5571
18	Employment (yrs)	364.41203	48.31750816	7.542029	0.001656	230.2608	498.5632	230.260841	498.563215
19	Experience (yrs)	227.61881	123.8361513	1.838064	0.139912	-116.206	571.4438	-116.20618	571.443799
20	Education (yrs)	266.93504	147.3556227	1.811502	0.144295	-142.191	676.0607	-142.1906	676.060686

◀

▶ Exercise 5 (pg. 482)	Finding a Multiple Regression Equation, the Standard Error of Estimate, and R^2

1. Open worksheet "ex9_4-5" in the Chapter 9 folder.

2. At the top of the screen, click **Tools → Data Analysis**. Select **Regression** and click **OK**.

If Data Analysis does not appear as a choice in the Tools menu, you will need to load the Microsoft Excel Analysis ToolPak add-in. Follow the procedure in Section GS 8.1 before continuing.

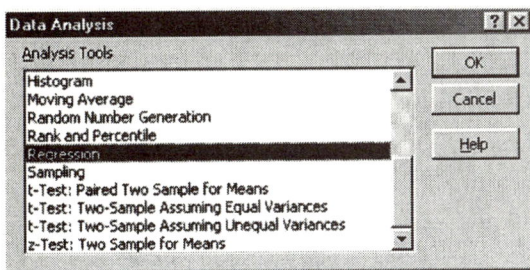

3.   Complete the Regression dialog box as shown below. Click **OK**.

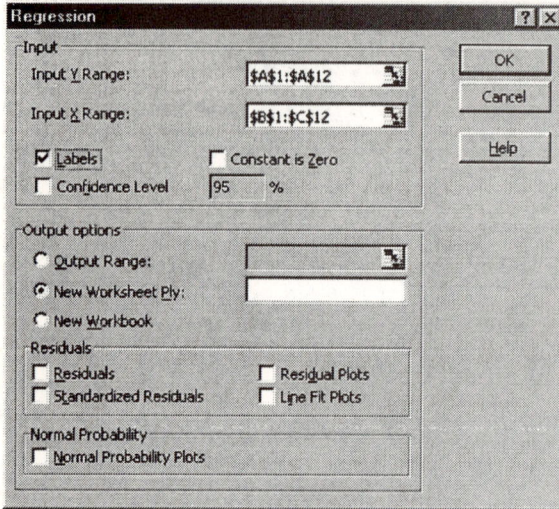

You will want to make some of the columns wider so that you can read all the labels. The coefficient of determination (R^2) is displayed in the upper portion of the output. R^2 is equal to 0.9881. The standard error of estimate is also displayed in the upper portion of the output. The standard error of estimate is equal to 34.1597. The coefficients for the multiple regression equation are displayed in the lower portion of the output under the label "Coefficients."

	A	B	C	D	E	F	G	H	I
1	SUMMARY OUTPUT								
2									
3	*Regression Statistics*								
4	Multiple R	0.9940231							
5	R Square	0.9880819							
6	Adjusted R Square	0.9851024							
7	Standard Error	34.159726							
8	Observations	11							
9									
10	ANOVA								
11		*df*	*SS*	*MS*	*F*	*ignificance F*			
12	Regression	2	773935.4867	386968	331.624	2.02E-08			
13	Residual	8	9335.095087	1166.89					
14	Total	10	783270.5818						
15									
16		*Coefficients*	*Standard Error*	*t Stat*	*P-value*	*Lower 95%*	*Upper 95%*	*Lower 95.0%*	*Upper 95.0%*
17	Intercept	-256.2926	44.46560248	-5.76384	0.000422	-358.831	-153.755	-358.83052	-153.75466
18	sq footage	103.50189	95.23861788	1.08676	0.308798	-116.119	323.1227	-116.11889	323.122683
19	Shopping centers	14.64888	10.9081471	1.34293	0.216148	-10.5054	39.80313	-10.505369	39.8031281

Technology Lab

| ▶ Exercise 1 (pg. 483) | Constructing a Scatter Plot |

1. Open worksheet "Tech9" in the Chapter 9 folder.

2. You will first construct a scatter plot of the tar and nicotine variables. Click on any cell within the table of data. At the top of the screen, click **Insert** and select **Chart** from the menu that appears.

3. In the Chart Type dialog box, select the **XY (Scatter)** chart type by clicking on it. Then click on the topmost chart sub-type. Click **Next**.

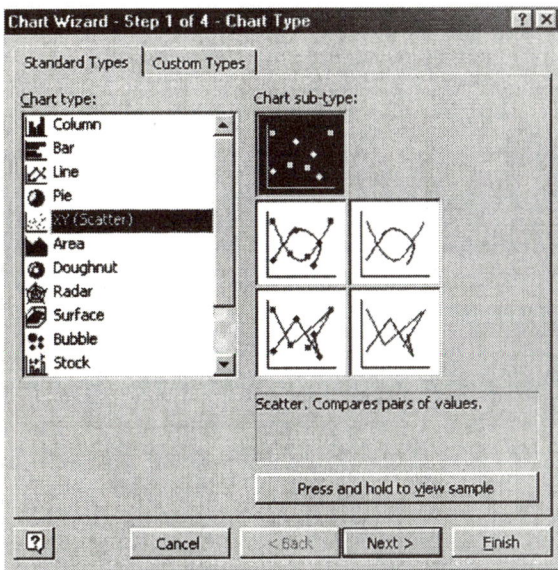

4. In the Chart Source Data dialog box, the entry in the data range field should be
    =Tech9!$B$1:$C$23. Correct the entry and then click **Next**.

    ---

    *Note that the data range includes only columns B and C. These are the columns that
    contain the tar and nicotine variables that will be included in this scatter plot.*

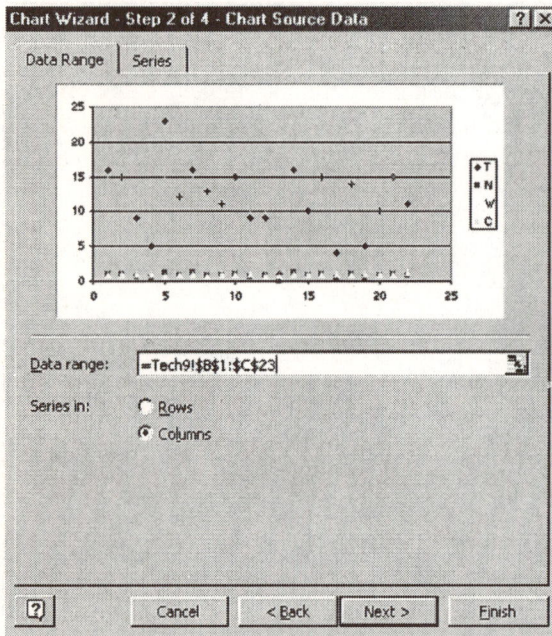

5. Click the Titles tab at the top of the Chart Options dialog box. In the Chart title
    field, change the title to **Tar by Nicotine Content of Cigarettes**. In the Value (X)
    axis field, enter **Tar in mg**. In the Value (Y) axis field, enter **Nicotine in mg**.

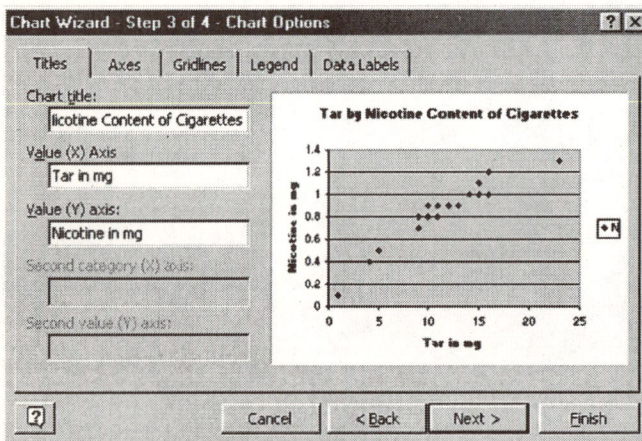

6. Click the Legend tab at the top of the Chart Options dialog box. Click in the box to the left of Show Legend so tat a checkmark does not appear there. This removes the N legend from the right side of the scatter plot. Click **Next**.

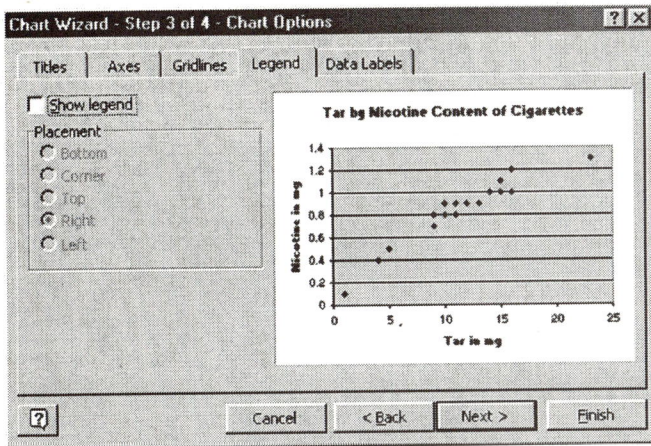

7. The Chart Location dialog box presents two options for placement of the scatter plot. For this example, select **As new sheet**. To do this, click the button to the left of **As new sheet** so that a black dot appears there. Click **Finish**.

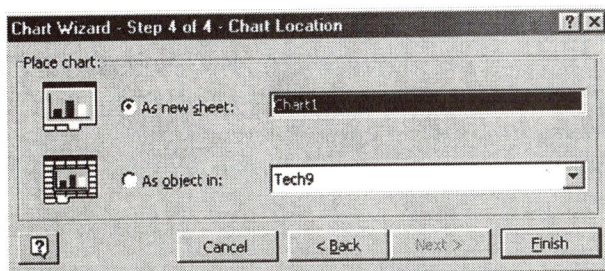

Your scatter plot should look similar to the scatter plot displayed below.

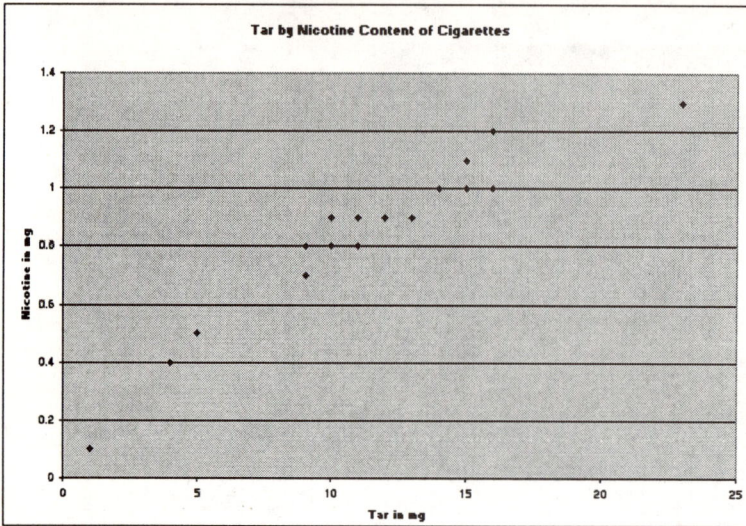

8. You will now construct the scatter plot for the tar and weight variables. Go back to the worksheet that contains the data. To do this, click on the **Tech9** tab near the bottom of the screen.

9. Click on any cell within the table of data. At the top of the screen, click **Insert** and select **Chart** from the menu that appears.

10. In the Chart Type dialog box, select the **XY (Scatter)** chart type by clicking on it. Then click on the topmost chart sub-type. Click **Next**.

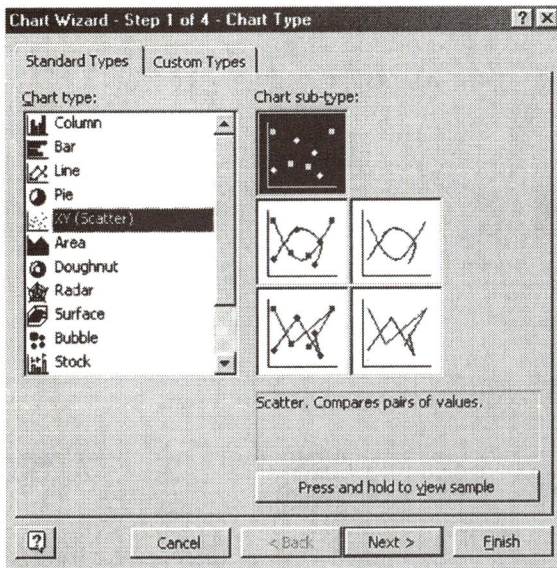

Chart Wizard - Step 1 of 4 - Chart Type

Standard Types | Custom Types

Chart type:
- Column
- Bar
- Line
- Pie
- XY (Scatter)
- Area
- Doughnut
- Radar
- Surface
- Bubble
- Stock

Chart sub-type:

Scatter. Compares pairs of values.

Press and hold to view sample

Cancel | < Back | Next > | Finish

11. Click the **Data Range** tab at the top of the Chart Source Data dialog box. Edit the entry in the data range field so that it reads **=Tech9!$B$1:$D$23**.

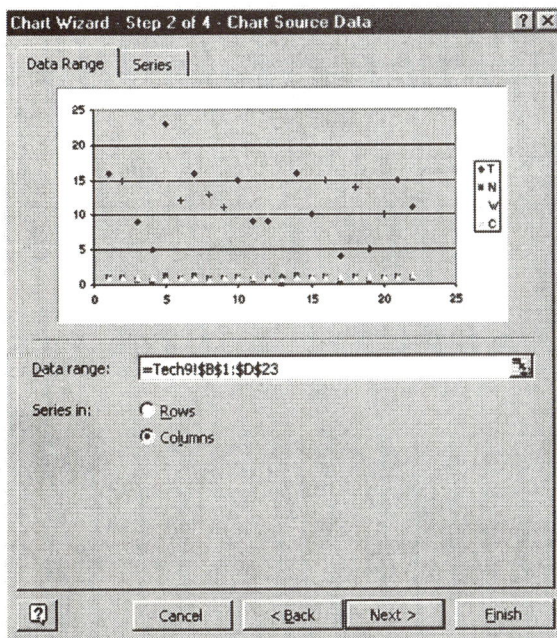

Chart Wizard - Step 2 of 4 - Chart Source Data

Data Range | Series

Data range: =Tech9!$B$1:$D$23

Series in: ○ Rows
● Columns

Cancel | < Back | Next > | Finish

12. Click the **Series** tab at the top of the Chart Source Data dialog box. For Y values, you want the weight variable that is in column D of the worksheet. Edit the entry in the Y Values field so that it reads **=Tech9!$D$2:$D$23**. Click **Next**.

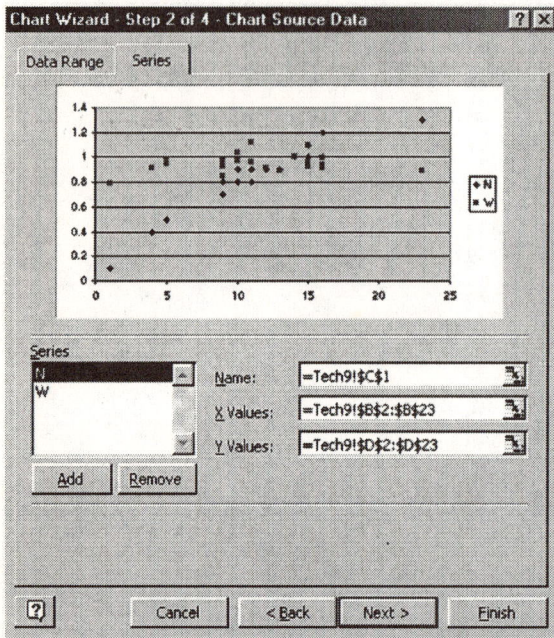

13. Click the Titles tab at the top of the Chart Options dialog box. In the Chart title field, enter **Tar Content by Weight of Cigarettes**. In the Value (X) axis field, enter **Tar in mg**. In the Value (Y) axis field, enter **Weight in g**.

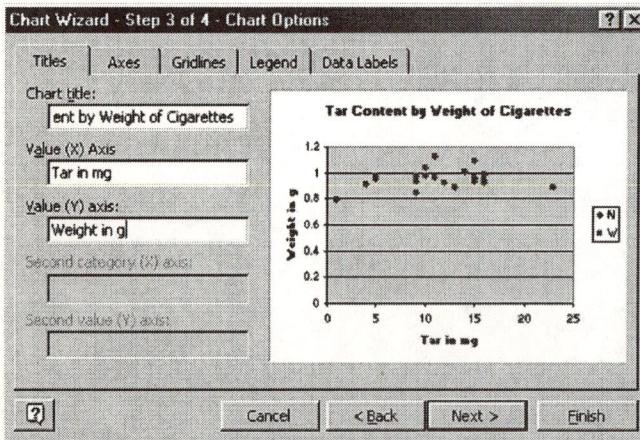

14. Click the **Legend** tab at the top of the Chart Options dialog box. Click in the box to the left of **Show Legend** so that a checkmark does not appear there. This removes the N and W legends from the right side of the scatter plot. Click **Next**.

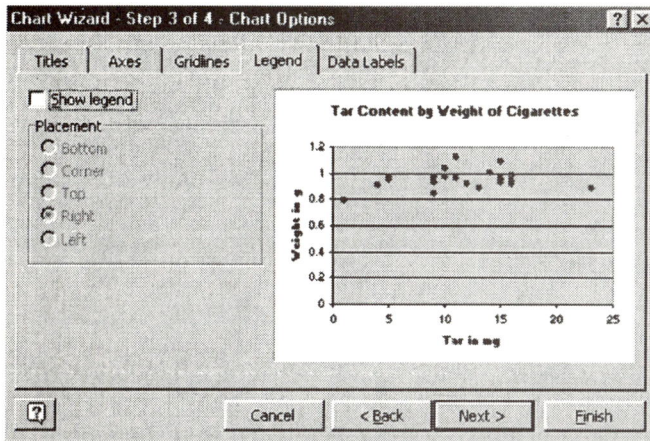

15. The Chart Location dialog box presents two options for placement of the scatter plot. Select **As new sheet**. Click **Finish**.

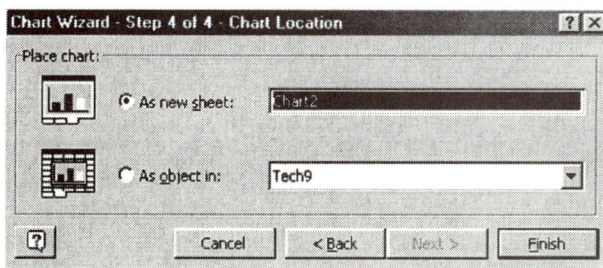

16. There is a great deal of unused space in the lower half of this scatter plot because the minimum weight value is 0.79. Instead of starting Y axis values with 0, start with 0.7. To do this, right click on any value along the right axis and select **Format Axis** from the shortcut menu that appears.

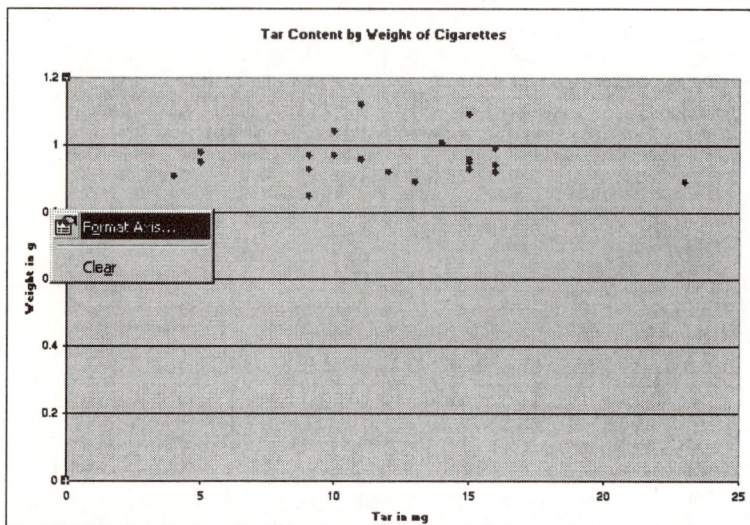

17. Click the **Scale** tab at the top of the Format Axis dialog box. Edit the entry in the Minimum window so that it reads **.7**. Click **OK**.

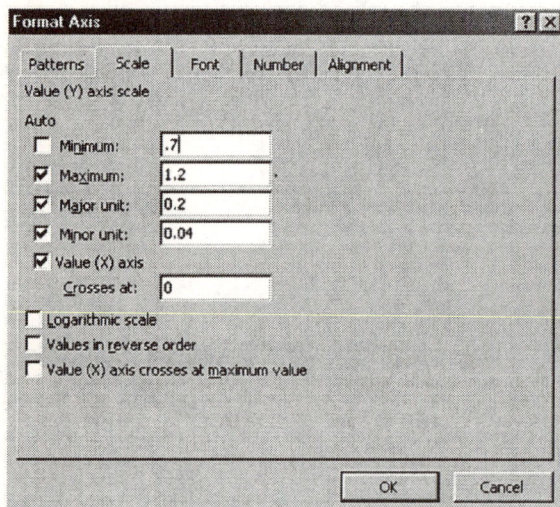

Your scatter plot should now look similar to the scatter plot shown below.

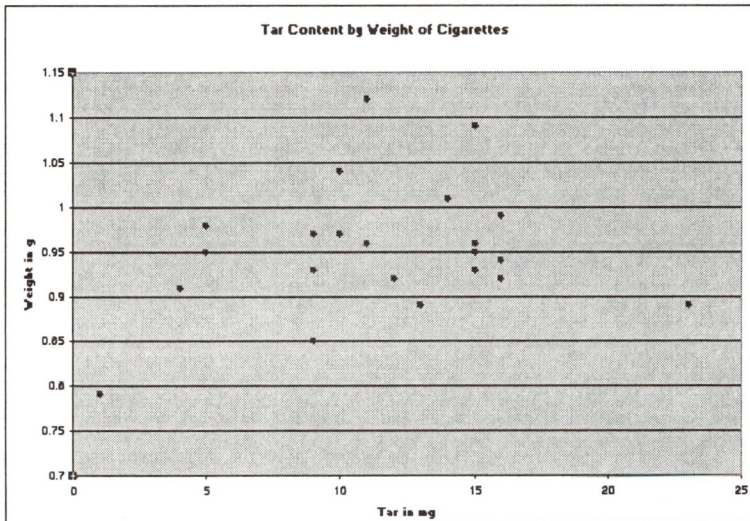

Tar Content by Weight of Cigarettes

▶ Exercise 3 (pg. 483) | Finding the Correlation Coefficient for Each Pair of Variables

1. Open worksheet "Tech9" in the Chapter 9 folder.

*If you have just completed Exercise 1 on page 483 and have not closed the Excel worksheet, return to the sheet containing the data by clicking on the **Tech9** tab at the bottom of the screen.*

2. At the top of the screen, click **Tools → Data Analysis**. Select **Correlation** and click **OK**.

If Data Analysis does not appear as a choice in the Tools menu, you will need to load the Microsoft Excel Analysis ToolPak add-in. Follow the procedure in Section GS 8.1 before continuing.

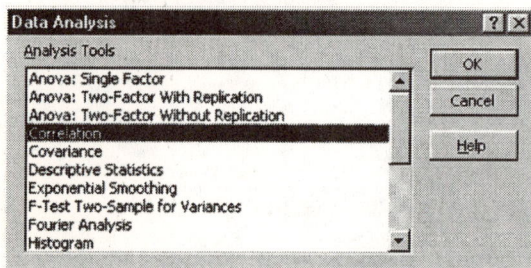

Data Analysis

Analysis Tools

Anova: Single Factor
Anova: Two-Factor With Replication
Anova: Two-Factor Without Replication
Correlation
Covariance
Descriptive Statistics
Exponential Smoothing
F-Test Two-Sample for Variances
Fourier Analysis
Histogram

OK   Cancel   Help

3. Complete the Correlation dialog box as shown below. Click **OK**.

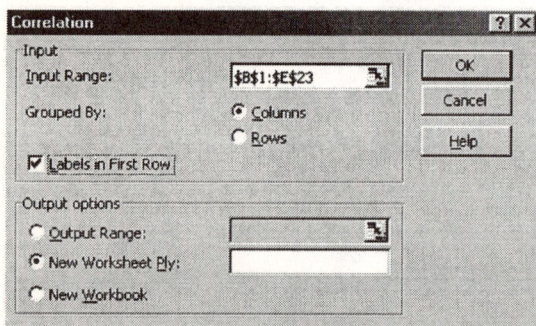

Correlation

Input

Input Range: $B$1:$E$23

Grouped By: ⊙ Columns
                ○ Rows

☑ Labels in First Row

Output options

○ Output Range:
⊙ New Worksheet Ply:
○ New Workbook

OK   Cancel   Help

The output is a correlation matrix that displays the correlation coefficients for all pairs of variables.

	A	B	C	D	E
1		T	N	W	C
2	T	1			
3	N	0.958782	1		
4	W	0.208178	0.298418	1	
5	C	0.91021	0.934994	0.361635	1

▶ **Exercise 4** (pg. 483)	Finding the Regression Line

1. Open worksheet "Tech9" in the Chapter 9 folder.

---

*If you have just completed Exercise 1 or Exercise 3 on page 483 and have not closed the Excel worksheet, return to the sheet containing the data by clicking on the **Tech9** tab at the bottom of the screen.*

---

2. You will first find the regression line equation for the tar and nicotine variables. Tar will be the Y variable in the equation and nicotine will be the X variable. At the top of the screen, click **Tools → Data Analysis**. Select **Regression** and click **OK**.

If Data Analysis does not appear as a choice in the Tools menu, you will need to load the Microsoft Excel Analysis ToolPak add-in. Follow the procedure in Section GS 8.1 before continuing.

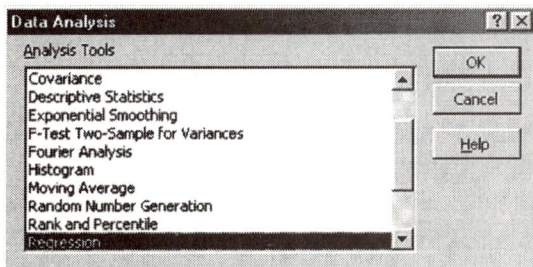

3. Complete the Regression dialog box as shown below. Click **OK**.

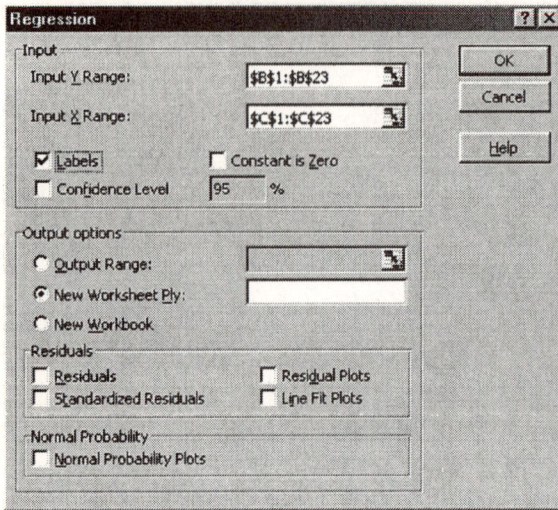

Make the columns in the output wider so that you can read all the labels. Your output should appear similar to the output shown below. The values for the intercept and slope of the regression line equation are shown in the lower part of the output. The intercept is –2.7149 and the slope is 16.6876.

	A	B	C	D	E	F	G	H	I
16		Coefficients	Standard Error	t Stat	P-value	Lower 95%	Upper 95%	Lower 95.0%	Upper 95.0%
17	Intercept	-2.714885	0.994704233	-2.72934	0.01292	-4.7898004	-0.63997	-4.7898004	-0.639969
18	N	16.687631	1.105847665	15.0904	2.15E-12	14.3808743	18.99439	14.3808743	18.9943878

4. You will now find the regression line equation for the tar and carbon monoxide variables. Tar will be the Y variable in the equation and carbon monoxide will be the X variable. Go back to the sheet containing the data by clicking on the **Tech9** tab at the bottom of the screen. Then, at the top of the screen, click **Tools → Data Analysis**. Select **Regression** and click **OK**.

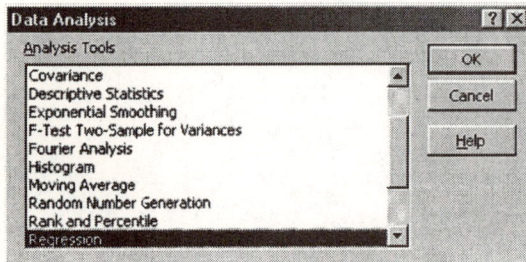

5. Complete the Regression dialog box as shown below. Click **OK**.

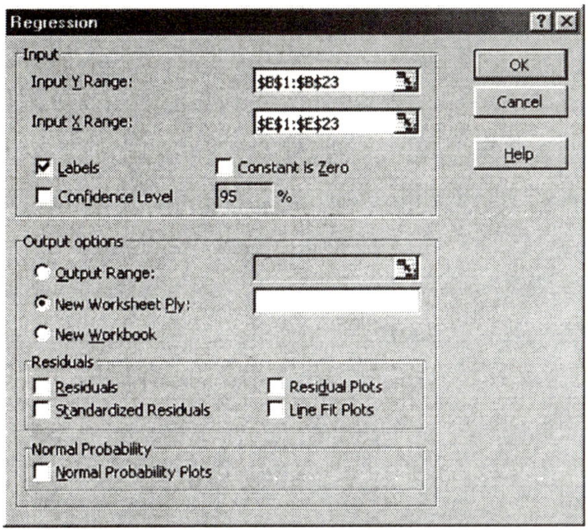

6. Make the columns in the output wider so that you can read all the labels. Your output should appear similar to the output shown below. The values for the intercept and slope of the regression line equation are shown in the lower part of the output. The intercept is −1.3882 and the slope is 1.1291.

	A	B	C	D	E	F	G	H	I
16		Coefficients	Standard Error	t Stat	P-value	Lower 95%	Upper 95%	Lower 95.0%	Upper 95.0%
17	Intercept	-1.388179	1.391541919	-0.99758	0.330399	-4.29088	1.514525	-4.2908831	1.51452532
18	C	1.1291267	0.114879059	9.82883	4.23E-09	0.889493	1.36876	0.88949332	1.36876014

◄

▶ Exercise 6 (pg. 483)   Finding Multiple Regression Equations

1. Open worksheet "Tech9" in the Chapter 9 folder.

*If you have just completed Exercise 1, Exercise 3, or Exercise 4 on page 483 and have not closed the Excel worksheet, return to the sheet containing the data by clicking on the **Tech9** tab at the bottom of the screen.*

2. At the top of the screen, click **Tools → Data Analysis**. Select **Regression** and click **OK**.

If Data Analysis does not appear as a choice in the Tools menu, you will need to load the Microsoft Excel Analysis ToolPak add-in. Follow the procedure in Section GS 8.1 before continuing.

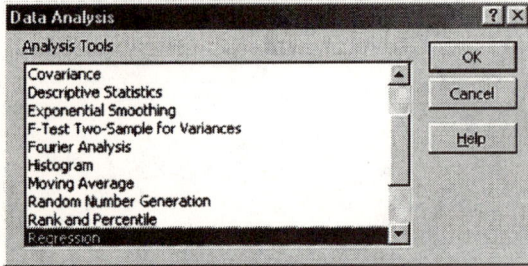

3. You will first construct the multiple regression equation using three predictor variables—nicotine, weight and carbon monoxide. Complete the Regression dialog box as shown below. Click **OK**.

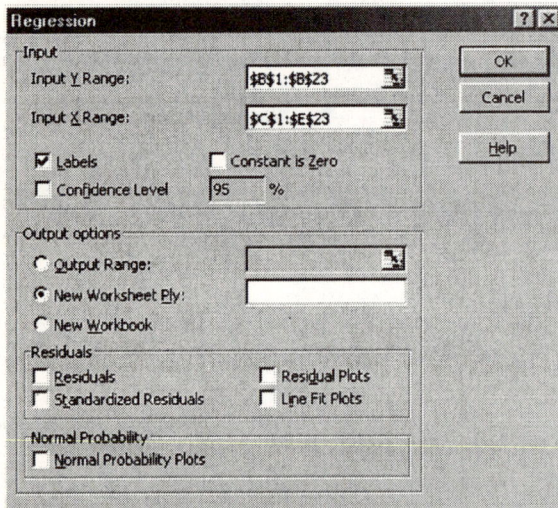

You will want to make some of the columns wider so that you can read all the labels. Your output should look similar to the output shown below. In the intercept and coefficients of the predictor variables are displayed in the bottom four lines of the output under the label "Coefficients."

	A	B	C	D	E	F	G	H	I
16		Coefficients	Standard Error	t Stat	P-value	Lower 95%	Upper 95%	Lower 95.0%	Upper 95.0%
17	Intercept	3.4469686	4.2381019	0.813328	0.426658	-5.45696	12.3509	-5.45696	12.3508972
18	N	14.350349	3.091521395	4.64184	0.000203	7.855298	20.8454	7.8552981	20.845399
19	W	-6.996539	4.656584728	-1.5025	0.150308	-16.7797	2.78659	-16.779668	2.78659032
20	C	0.2183635	0.225569488	0.968054	0.345846	-0.25554	0.692268	-0.2555408	0.69226773

4. Now you will construct the regression equation using two predictor variables—nicotine and carbon monoxide. Return to the worksheet containing the data by clicking on the **Tech9** tab at the bottom of the screen.

5. The predictor variables must be located in adjacent columns in the Excel worksheet. So you'll begin by making a copy of the worksheet and then deleting the column of data between nicotine and carbon monoxide. At the top of the screen, click **Edit → Move or Copy Sheet**.

6. In the Move or Copy dialog box, click in the box to the left of **Create a copy** to place a checkmark there. Click **OK**.

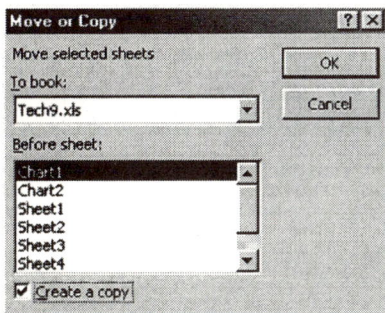

7. The copy is called "Tech9(2)." To delete the column with the weight data, first click directly on ▢ D ▢ in the letter row at the top of the worksheet. Column D will be highlighted. Then, at the top of the screen, click **Edit → Delete**. Your worksheet should now look like the one shown below.

	A	B	C	D	E	F	G
1	Brand	T	N	C			
2	Alpine	16	1	15			
3	Benson &	15	1.1	15			
4	Camel	9	0.7	11			
5	Carlton	5	0.5	3			
6	Chesterfiel	23	1.3	15			
7	Kent	12	0.9	12			

8. At the top of the screen, click **Tools → Data Analysis**. Select **Regression** and click **OK**.

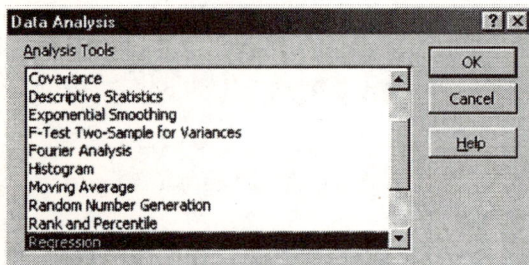

Data Analysis   [?][X]

Analysis Tools

Covariance
Descriptive Statistics
Exponential Smoothing
F-Test Two-Sample for Variances
Fourier Analysis
Histogram
Moving Average
Random Number Generation
Rank and Percentile
Regression

OK
Cancel
Help

9. Complete the Regression dialog box as shown below. Click **OK**.

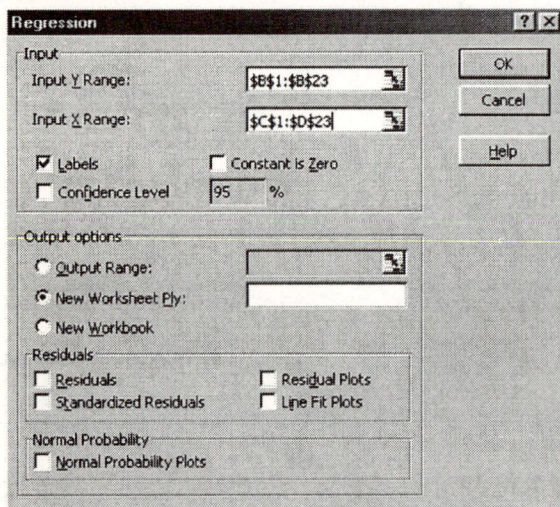

Regression   [?][X]

Input

Input Y Range: $B$1:$B$23

Input X Range: $C$1:$D$23

☑ Labels   ☐ Constant is Zero
☐ Confidence Level   95 %

OK
Cancel
Help

Output options

☐ Output Range:
☉ New Worksheet Ply:
☐ New Workbook

Residuals

☐ Residuals   ☐ Residual Plots
☐ Standardized Residuals   ☐ Line Fit Plots

Normal Probability

☐ Normal Probability Plots

After making some of the columns wider, your output should appear similar to the output shown below. The intercept and coefficients of the predictor variables are displayed in the bottom lines of the output under the label "Coefficients."

	A	B	C	D	E	F	G	H	I
16		Coefficients	Standard Error	t Stat	P-value	Lower 95%	Upper 95%	Lower 95.0%	Upper 95.0%
17	Intercept	-2.748014	1.012498867	-2.71409	0.013764	-4.8672	-0.62883	-4.8671992	-0.6288289
18	N	14.908175	3.16908666	4.70425	0.000154	8.275198	21.54115	8.2751979	21.5411513
19	C	0.1356453	0.225871018	0.600543	0.555241	-0.33711	0.608399	-0.3371083	0.60839896

Chi-Square Tests and the F-Distribution

CHAPTER 10

Section 10.2

▸ Example 3 (pg. 509) — Using Technology for a Chi-Square Independence Test

If the PHStat add-in has not been loaded, you will need to load it before continuing. Follow the instructions in Section GS 8.2.

1. First, open a new Excel worksheet. Then, at the top of the screen, select **PHStat → Multiple-Sample Tests → Chi-Square Test**.

2. Complete the Chi-Square Test dialog box as shown below. Click **OK**.

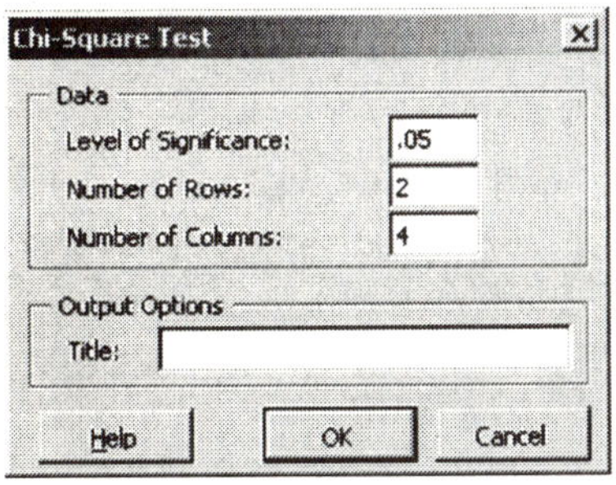

3. A Chi-Square Test worksheet will appear. Complete the Observed Frequencies table in this worksheet as shown below. You will notice that values in other locations of the worksheet change as you enter the observed frequencies.

	A	B	C	D	E	F
1	Chi-Square Test					
2						
3			Observed Frequencies			
4			Column variable			
5	Row variable	C1	C2	C3	C4	Total
6	R1	40	53	26	6	125
7	R2	34	68	37	11	150
8	Total	74	121	63	17	275

The critical value, the obtained chi-square test statistic, and the p-value are displayed in rows 24 through 26. The statistical decision is displayed in row 27.

17	Data	
18	Level of Significance	0.05
19	Number of Rows	2
20	Number of Columns	4
21	Degrees of Freedom	3
22		
23	Results	
24	Critical Value	7.814725
25	Chi-Square Test Statistic	3.493357
26	p-Value	0.321625
27	Do not reject the null hypothesis	
28		
29	Expected frequency assumption	
30	is met.	

▶ Exercise 5 (pg. 511)   Testing the Drug for Treatment of Obsessive-Compulsive Disorder

If the PHStat add-in has not been loaded, you will need to load it before continuing. Follow the instructions in Section GS 8.2.

1. Open a new Excel worksheet. At the top of the screen, select **PHStat → Multiple-Sample Tests → Chi-Square Test**.

2. Complete the Chi-Square Test dialog box as shown below. Click **OK**.

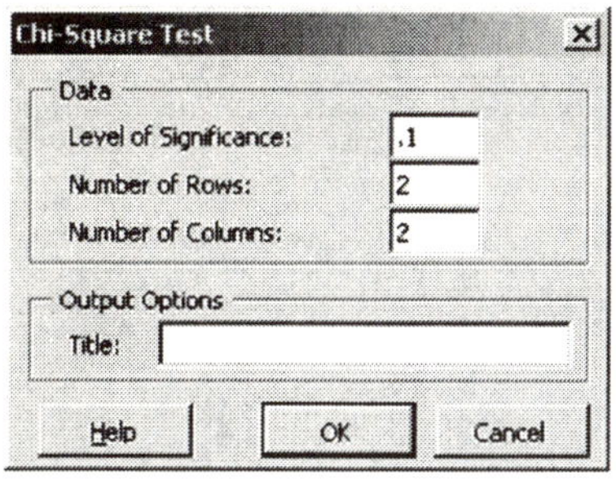

```
Chi-Square Test                                    [x]
┌─ Data ───────────────────────────────────────┐
│   Level of Significance:      [.1    ]        │
│   Number of Rows:             [2     ]        │
│   Number of Columns:          [2     ]        │
└───────────────────────────────────────────────┘
┌─ Output Options ─────────────────────────────┐
│   Title:  [                              ]    │
└───────────────────────────────────────────────┘
      [ Help ]      [   OK   ]      [ Cancel ]
```

4. A Chi-Square Test worksheet will appear. Complete the Observed Frequencies table in this worksheet as shown below. Values in other locations of the worksheet change as you enter the observed frequencies.

	A	B	C	D
1	Chi-Square Test			
2				
3	Observed Frequencies			
4		Column variable		
5	Row variable	C1	C2	Total
6	R1	39	25	64
7	R2	54	70	124
8	Total	93	95	188

The critical value, the obtained chi-square test statistic, and the p-value are displayed in rows 24 through 26. The statistical decision is displayed in row 27.

17	Data	
18	Level of Significance	0.1
19	Number of Rows	2
20	Number of Columns	2
21	Degrees of Freedom	1
22		
23	Results	
24	Critical Value	2.705541
25	Chi-Square Test Statistic	5.106317
26	p-Value	0.023839
27	Reject the null hypothesis	
28		
29	*Expected frequency assumption*	
30	*is met.*	

Section 10.3

> ► **Example 3** (pg. 519) | Performing a Two-Sample F-Test

If the PHStat add-in has not been loaded, you will need to load it before continuing. Follow the instructions in Section GS 8.2.

1. First open a new Excel worksheet. Then, at the top of the screen, select **PHStat →
Two-Sample Tests → F Test for Differences in Two Variances**.

2. Complete the F Test for Differences in Two Variances dialog box as shown below.
Click **OK**.

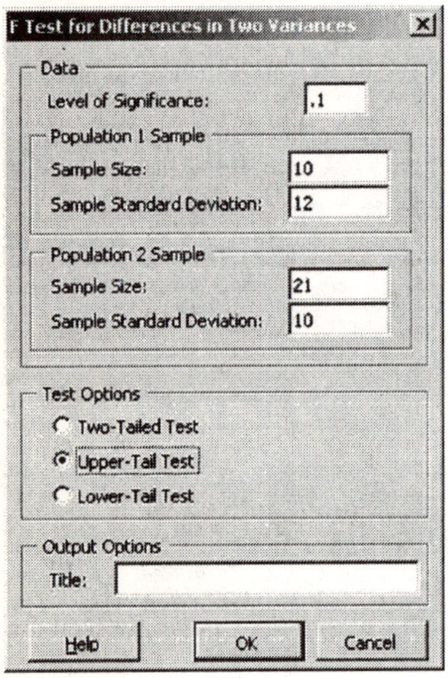

F Test for Differences in Two Variances	⊠
Data	
Level of Significance:	.1
Population 1 Sample	
Sample Size:	10
Sample Standard Deviation:	12
Population 2 Sample	
Sample Size:	21
Sample Standard Deviation:	10
Test Options	
○ Two-Tailed Test	
● Upper-Tail Test	
○ Lower-Tail Test	
Output Options	
Title:	
Help	OK	Cancel

Your output should appear similar to the output displayed below.

	A	B	C	D	E
1	F Test for Differences in Two Variances				
2					
3	Data				
4	Level of Significance	0.1			
5	Population 1 Sample				
6	Sample Size	10			
7	Sample Standard Deviation	12			
8	Population 2 Sample				
9	Sample Size	21			
10	Sample Standard Deviation	10			
11					
12	Intermediate Calculations				
13	F-Test Statistic	1.44			
14	Population 1 Sample Degrees of Freedom	9			
15	Population 2 Sample Degrees of Freedom	20			
16				Calculations Area	
17				FDIST value	0.2369
18	Upper-Tail Test			1-FDIST value	0.7631
19	Upper Critical Value	1.964853			
20	p-Value	0.2369			
21	Do not reject the null hypothesis				

◀

Section 10.4

▶ Example 2 (pg. 529)   Using Technology to Perform ANOVA Tests

1. Open worksheet "Airline" in the Chapter 10 folder.

2.   At the top of the screen, click **Tools** → **Data Analysis**. Select **ANOVA: Single Factor** and click **OK**.

If Data Analysis does not appear as a choice in the Tools menu, you will need to load the Microsoft Excel Analysis ToolPak add-in. Follow the procedure in Section GS 8.1 before continuing.

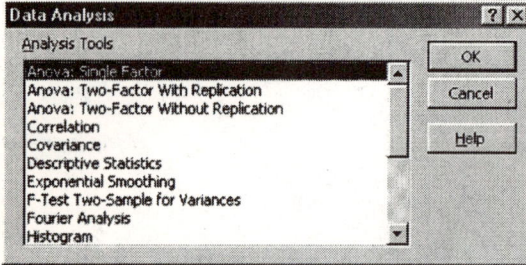

3.   Complete the ANOVA: Single Factor dialog box as shown below. Click **OK**.

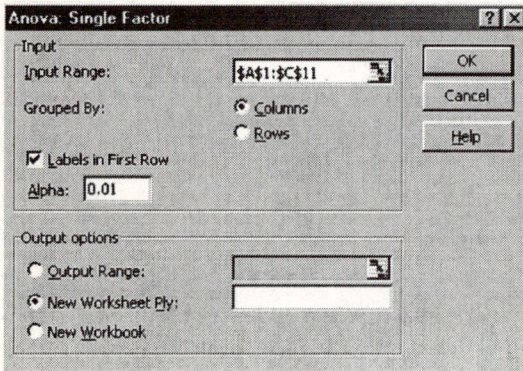

Make column A wider so that you can read all the labels. Your output should appear similar to the output displayed below.

	A	B	C	D	E	F	G
1	Anova: Single Factor						
2							
3	SUMMARY						
4	*Groups*	*Count*	*Sum*	*Average*	*Variance*		
5	Airline1	10	1238	123.8	106.6222		
6	Airline2	10	1315	131.5	120.2778		
7	Airline3	10	1427	142.7	215.1222		
8							
9							
10	ANOVA						
11	*Source of Variation*	*SS*	*df*	*MS*	*F*	*P-value*	*F crit*
12	Between Groups	1806.467	2	903.2333	6.130235	0.006383	5.488118
13	Within Groups	3978.2	27	147.3407			
14							
15	Total	5784.667	29				

◀

▶ Exercise 1 (pg. 531)	Testing the Claim That the Mean Toothpaste Costs Per Month are Different

1. Open worksheet "Ex10_4-1" in the Chapter 10 folder.

2. At the top of the screen, click **Tools → Data Analysis**. Select **ANOVA: Single Factor** and click **OK**.

If Data Analysis does not appear as a choice in the Tools menu, you will need to load the Microsoft Excel Analysis ToolPak add-in. Follow the procedure in Section GS 8.1 before continuing.

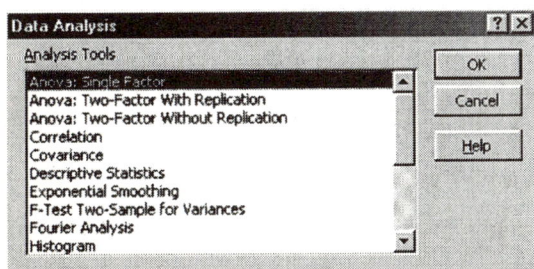

3.   Complete the ANOVA: Single Factor dialog box as shown below. Click **OK**.

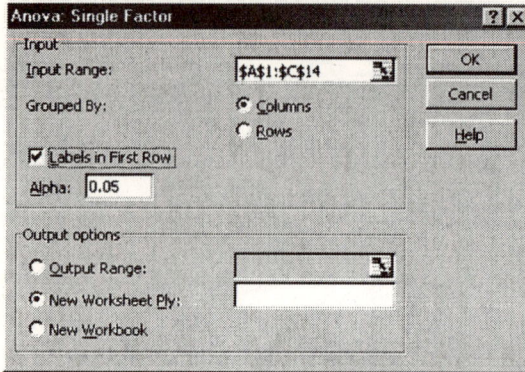

Anova: Single Factor		? X
Input		
Input Range:	$A$1:$C$14	OK
Grouped By:	⊙ Columns	Cancel
	○ Rows	Help
☑ Labels in First Row		
Alpha: 0.05		
Output options		
○ Output Range:		
⊙ New Worksheet Ply:		
○ New Workbook		

Make column A wider so that you can read all the labels. Your output should appear similar to the output displayed below.

	A	B	C	D	E	F	G
1	Anova: Single Factor						
2							
3	SUMMARY						
4	Groups	Count	Sum	Average	Variance		
5	Moderate	13	15.49	1.191538	0.677964		
6	Low	13	12.52	0.963077	0.145173		
7	Very Low	4	6.75	1.6875	2.530358		
8							
9							
10	ANOVA						
11	Source of Variation	SS	df	MS	F	P-value	F crit
12	Between Groups	1.630026	2	0.815013	1.2597	0.299888	3.354131
13	Within Groups	17.46872	27	0.64699			
14							
15	Total	19.09875	29				

Technology Lab

▶ Exercise 3 (pg. 536)	Determining Whether the Populations Have Equal Variances

If the PHStat add-in has not been loaded, you will need to load it before continuing. Follow the instructions in Section GS 8.2.

1. Open worksheet "Tech10-a" in the Chapter 10 folder.

2. Because you are working with raw data rather than summary statistics, you first need to calculate the standard deviations of the three groups. At the top of the screen, click **Tools → Data Analysis**. Select **Descriptive Statistics** and click **OK**.

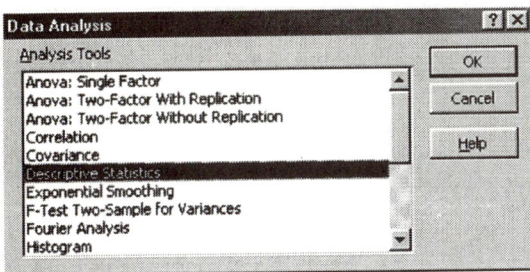

3. Complete the Descriptive Statistics dialog box as shown below. Click **OK**.

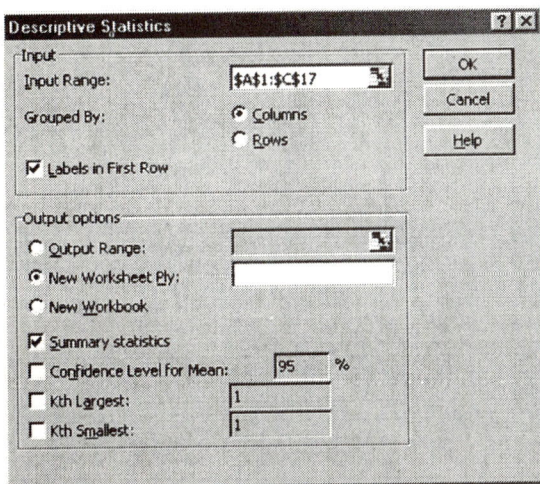

4. Make the columns wider so that you can read the labels. The standard deviations of medium, heavy, and multiple purpose vehicles are 436.46, 448.43, and 566.66, respectively. For this exercise, instructions are provided only for testing the equality of the medium and heavy vehicles. At the top of the screen, select **PHStat → Two-Sample Tests → F Test for Differences in Two Variances**.

	A	B	C	D	E	F
1	Medium		Heavy		Multiple	
2						
3	Mean	1042.75	Mean	947.9375	Mean	971.125
4	Standard Error	109.116	Standard Error	112.1066	Standard Error	141.6639
5	Median	978.5	Median	872.5	Median	966.5
6	Mode	#N/A	Mode	#N/A	Mode	430
7	Standard Deviation	436.4639	Standard Deviation	448.4264	Standard Deviation	566.6558
8	Sample Variance	190500.7	Sample Variance	201086.2	Sample Variance	321098.8
9	Kurtosis	-1.43386	Kurtosis	-0.92956	Kurtosis	-1.24662
10	Skewness	0.001657	Skewness	0.107502	Skewness	0.384132
11	Range	1325	Range	1468	Range	1644
12	Minimum	332	Minimum	260	Minimum	267
13	Maximum	1657	Maximum	1728	Maximum	1911
14	Sum	16684	Sum	15167	Sum	15538
15	Count	16	Count	16	Count	16

5. Complete the F Test for Differences in Two Variances dialog box as shown below. Click **OK**.

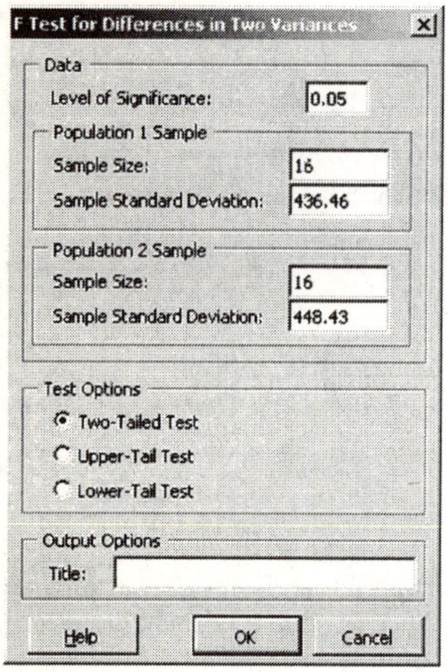

Your output should appear similar to the output displayed below.

	A	B	C	D	E
1	F Test for Differences in Two Variances				
2					
3	Data				
4	Level of Significance	0.05			
5	Population 1 Sample				
6	Sample Size	16			
7	Sample Standard Deviation	436.46			
8	Population 2 Sample				
9	Sample Size	16			
10	Sample Standard Deviation	448.43			
11					
12	Intermediate Calculations				
13	F-Test Statistic	0.947326			
14	Population 1 Sample Degrees of Freedom	15			
15	Population 2 Sample Degrees of Freedom	15			
16				Calculations Area	
17	Two-Tailed Test			FDIST value	0.541038
18	Lower Critical Value	0.349395		1-FDIST value	0.458962
19	Upper Critical Value	2.862095			
20	p-Value	0.917923			
21	Do not reject the null hypothesis				

◀

▶ Exercise 4 (pg. 536)

Testing the Claim That the Three Types of Vehicles Have the Same Injury Potential

1. Open worksheet "Tech10-a" in the Chapter 10 folder.

---

*If you have just completed Exercise 3 on page 536 and have not yet closed the Excel worksheet, return to the sheet with the data by clicking on the **Tech10-a** tab near the bottom of the screen.*

---

2. At the top of the screen, click **Tools → Data Analysis**. Select **ANOVA: Single Factor** and click **OK**.

If Data Analysis does not appear as a choice in the Tools menu, you will need to load the Microsoft Excel Analysis ToolPak add-in. Follow the procedure in Section GS 8.1 before continuing.

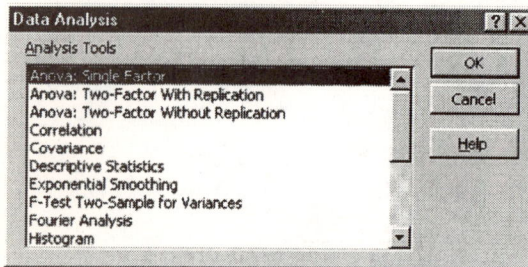

3. Complete the ANOVA: Single Factor dialog box as shown below. Click **OK**.

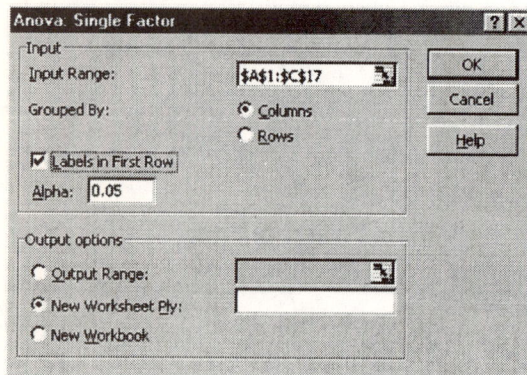

Make column A wider so that you can read all the labels. Your output should appear similar to the output shown below.

	A	B	C	D	E	F	G
1	Anova: Single Factor						
2							
3	SUMMARY						
4	Groups	Count	Sum	Average	Variance		
5	Medium	16	16684	1042.75	190500.7		
6	Heavy	16	15167	947.9375	201086.2		
7	Multiple	16	15538	971.125	321098.8		
8							
9							
10	ANOVA						
11	Source of Variation	SS	df	MS	F	P-value	F crit
12	Between Groups	78171.79	2	39085.9	0.164529	0.848801	3.20432
13	Within Groups	10690286	45	237561.9			
14							
15	Total	10768457	47				

Nonparametric Tests

Section 11.2

▶ Example 2 (pg. 561)	Performing a Wilcoxon Rank Sum Test

If the PHStat add-in has not been loaded, you will need to load it before continuing. Follow the instructions in Section GS 8.2.

1. Open worksheet "earnings." in the Chapter 11 folder.

2. At the top of the screen, select **PHStat → Two-Sample Tests → Wilcoxon Rank Sum Test**.

3. Complete the Wilcoxon Rank Sum Test dialog box as shown below. The Population 1 Sample Cell Range is A1 through A11. The Population 2 Sample Cell Range is B1 through B13. Click **OK**.

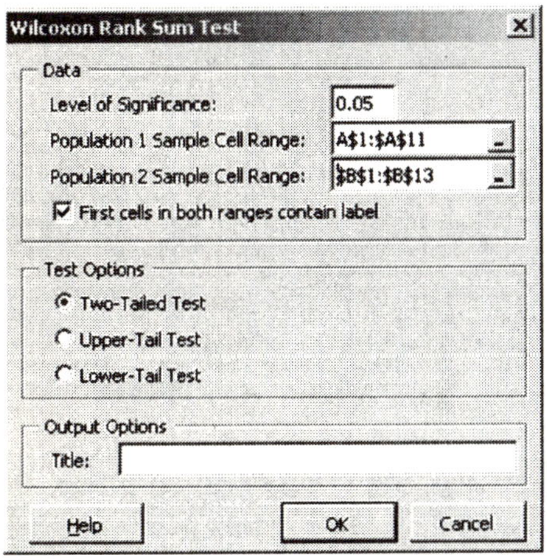

Your output should look similar to the output displayed below.

	A	B
1	Wilcoxon Rank Sum Test	
2		
3	Data	
4	Level of Significance	0.05
5		
6	Population 1 Sample	
7	Sample Size	10
8	Sum of Ranks	138
9	Population 2 Sample	
10	Sample Size	12
11	Sum of Ranks	115
12		
13	Intermediate Calculations	
14	Total Sample Size n	22
15	T1 Test Statistic	138
16	T1 Mean	115
17	Standard Error of T1	15.16575
18	Z Test Statistic	1.516575
19		
20	Two-Tailed Test	
21	Lower Critical Value	-1.95996
22	Upper Critical Value	1.959961
23	p-value	0.129374
24	Do not reject the null hypothesis	

◀

Section 11.3

▶ Example 1 (pg. 569)	Performing a Kruskal-Wallis Rank Test

If the PHStat add-in has not been loaded, you will need to load it before continuing. Follow the instructions in Section GS 8.2.

1. Open worksheet "payrates" in the Chapter 11 folder.

2. At the top of the screen, select **PHStat → Multiple-Sample Tests → Kruskal-Wallis Rank Test**.

3. Complete the Kruskal-Wallis Rank Test dialog box as shown below. The Sample Data Cell Range is A1 through C11. Click **OK**.

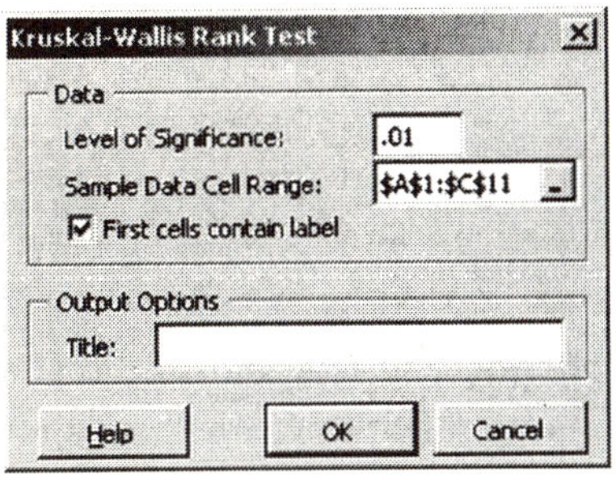

Kruskal-Wallis Rank Test	×
Data	
Level of Significance:	.01
Sample Data Cell Range:	$A$1:$C$11 ▪
☑ First cells contain label	
Output Options	
Title:	
Help	OK	Cancel

4. Your output should appear similar to the output shown below. This output is not completely accurate. You will notice that it shows a sample size of 10 for Group 2, and it should be a sample size of 9. There is a small error in the computer program that occurs when sample sizes are not equal. However, it is simple for you to fix.

	A	B
1	Kruskal-Wallis Rank Test	
2		
3	Data	
4	Level of Significance	0.01
5		
6	Group 1	
7	Sum of Ranks	94.5
8	Sample Size	10
9	Group 2	
10	Sum of Ranks	223
11	Sample Size	10
12	Group 3	
13	Sum of Ranks	147.5
14	Sample Size	10
15		
16	Intermediate Calculations	
17	Sum of Squared Ranks/Sample Size	8041.55
18	Sum of Sample Sizes	30
19	Number of groups	3
20	H Test Statistic	10.76194
21		
22	Test Result	
23	Critical Value	9.210351
24	p-Value	0.004603
25	Reject the null hypothesis	

5. Click in cell **B7** where you see 94.5. In the formula bar, you will see
    =SUMIF(Sorted!A2:A31,"MI",Sorted!C2:C31). This is a function referring to
    ranks displayed in the worksheet named Sorted. For this problem, it should say A30
    instead of A31. So, up in the formula bar, change A31 to A30.

CHIDIST	▾	X ✓ =	=SUMIF(Sorted!A2:A30,"MI",Sorted!C2:C31)

6. Next, click in cell **B8** where you see a sample size of 10. In the formula bar, you
    will see =COUNTIF(Sorted!A2 :A31,"MI"). Change A31 to A30 here also.

CHIDIST	▾	X ✓ =	=COUNTIF(Sorted!A2:A30,"MI")

7. Make the same changes in the SUMIF and COUNTIF functions for Group 2 and
    Group 3. These changes will cause numbers in the lower part of the output to
    change to correct values. The corrected output is shown below.

	A	B
1	**Kruskal-Wallis Rank Test**	
2		
3	**Data**	
4	**Level of Significance**	**0.01**
5		
6	Group 1	
7	Sum of Ranks	94.5
8	Sample Size	10
9	Group 2	
10	Sum of Ranks	193
11	Sample Size	9
12	Group 3	
13	Sum of Ranks	147.5
14	Sample Size	10
15		
16	Intermediate Calculations	
17	Sum of Squared Ranks/Sample Size	7207.428
18	Sum of Sample Sizes	29
19	Number of groups	3
20	H Test Statistic	9.412797
21		
22	**Test Result**	
23	**Critical Value**	9.210351
24	p-Value	0.009037
25	**Reject the null hypothesis**	

Technology Lab

▶ Exercise 3 (pg. 579)	Performing a Wilcoxon Rank Sum Test

If the PHStat add-in has not been loaded, you will need to load it before continuing. Follow the instructions in Section GS 8.2.

1. Open worksheet "Tech11_a" in the Chapter 11 folder.

2. At the top of the screen, select **PHStat → Two-Sample Tests → Wilcoxon Rank Sum Test**.

3. Complete the Wilcoxon Rank Sum Test dialog box as shown below. The Population 1 Sample Cell Range is A1 through A13, and the Population 2 Sample Cell Range is B1 through B13. Click **OK**.

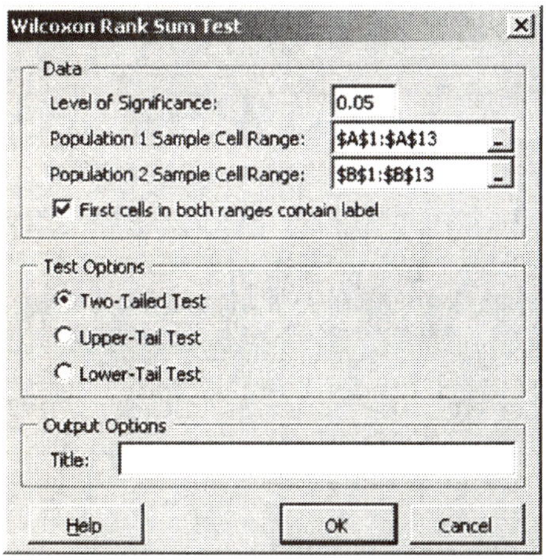

Your output should look similar to the output shown below.

	A	B
1	**Wilcoxon Rank Sum Test**	
2		
3	**Data**	
4	**Level of Significance**	0.05
5		
6	Population 1 Sample	
7	Sample Size	12
8	Sum of Ranks	219
9	Population 2 Sample	
10	Sample Size	12
11	Sum of Ranks	81
12		
13	Intermediate Calculations	
14	Total Sample Size n	24
15	T1 Test Statistic	219
16	T1 Mean	150
17	Standard Error of T1	17.32051
18	Z Test Statistic	3.983717
19		
20	**Two-Tailed Test**	
21	**Lower Critical Value**	-1.95996
22	**Upper Critical Value**	1.959961
23	***p*-value**	6.79E-05
24	**Reject the null hypothesis**	

◄

► **Exercise 4 (pg. 579)**   Performing a Kruskal-Wallis Rank Test

---

If the PHStat add-in has not been loaded, you will need to load it before continuing. Follow the instructions in Section GS 8.2.

---

1. Open worksheet "Tech11_a" in the Chapter 11 folder.

---

*If you have just completed Exercise 3 on page 579 and have not yet closed the worksheet, click on the **Tech11_a** tab at the bottom of the screen to return to the worksheet containing the data.*

---

2. At the top of the screen, select **PHStat → Multiple-Sample Tests → Kruskal-Wallis Rank Test**.

3. Complete the Kruskal-Wallis Rank Test dialog box as shown below. The Sample
    Data Cell Range is A1 through D13. Click **OK**.

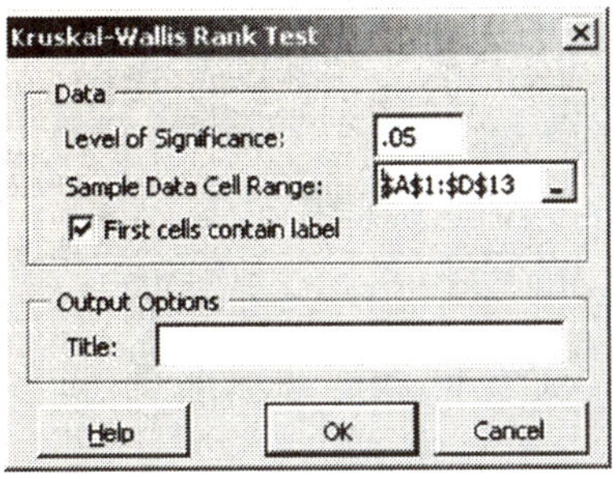

Kruskal-Wallis Rank Test	✕
Data	
Level of Significance:	.05
Sample Data Cell Range:	$A$1:$D$13
☑ First cells contain label	
Output Options	
Title:	
Help	OK	Cancel

Your output should appear similar to the output shown below. You may have to scroll
down a couple rows to see the entire output.

	A	B
1	**Kruskal-Wallis Rank Test**	
2		
3	**Data**	
4	**Level of Significance**	**0.05**
5		
6	Group 1	
7	Sum of Ranks	327.5
8	Sample Size	12
9	Group 2	
10	Sum of Ranks	116
11	Sample Size	12
12	Group 3	
13	Sum of Ranks	224.5
14	Sample Size	12
15	Group 4	
16	Sum of Ranks	508
17	Sample Size	12
18		
19	Intermediate Calculations	
20	Sum of Squared Ranks/Sample Size	35764.71
21	Sum of Sample Sizes	48
22	Number of groups	4
23	H Test Statistic	35.473
24		
25	**Test Result**	

◀

▶ Exercise 5 (pg. 579)	Performing a One-Way ANOVA

1.   Open worksheet "Tech11_a" in the Chapter 11 folder.

> *If you have just completed Exercise 3 or Exercise 4 on page 579 and have not yet closed the worksheet, click on the **Tech11_a** tab at the bottom of the screen to return to the worksheet containing the data.*

2.   At the top of the screen, click **Tools → Data Analysis**. Select **Anova: Single Factor** and click **OK**.

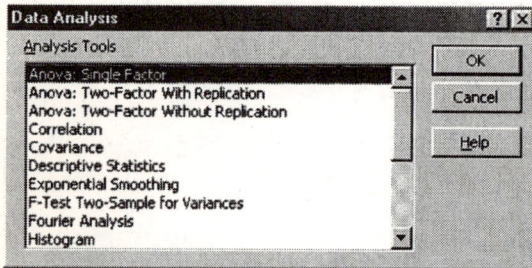

3.   Complete the Anova: Single Factor dialog box as shown below. Click **OK**.

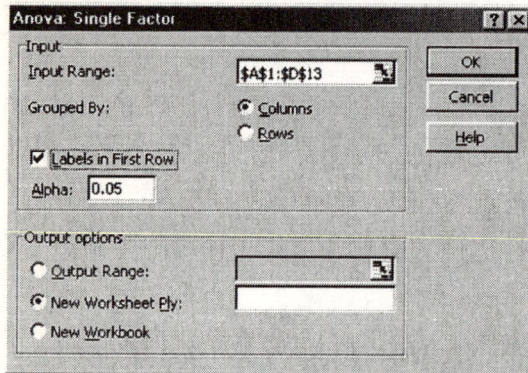

Adjust the column width so that you can read all the labels. Your output should appear similar to the output shown below.

	A	B	C	D	E	F	G
1	Anova: Single Factor						
2							
3	SUMMARY						
4	Groups	Count	Sum	Average	Variance		
5	NE	12	1692.7	141.0583	72.0772		
6	MW	12	1489.4	124.1167	58.58697		
7	S	12	1594.4	132.8667	86.59697		
8	W	12	2093.8	174.4833	249.8652		
9							
10							
11	ANOVA						
12	Source of Variation	SS	df	MS	F	P-value	F crit
13	Between Groups	17449.99	3	5816.665	49.80807	3.39E-14	2.816464
14	Within Groups	5138.389	44	116.7816			
15							
16	Total	22588.38	47				